I0056578

Materials and Contact Characterisation X

WITPRESS

WIT Press publishes leading books in Science and Technology.
Visit our website for new and current list of titles.
www.witpress.com

WITeLibrary

Home of the Transactions of the Wessex Institute.
Papers published in this volume are archived in the WIT eLibrary in volume 133 of
WIT Transactions on Engineering Sciences (ISSN 1743-3533).
The WIT eLibrary provides the international scientific community with immediate
and permanent access to individual papers presented at WIT conferences.
http://library.witpress.com

TENTH INTERNATIONAL CONFERENCE ON
COMPUTATIONAL METHODS AND EXPERIMENTS IN
MATERIAL AND CONTACT CHARACTERISATION

MATERIALS CHARACTERISATION 2021

CONFERENCE CHAIRMEN

Santiago Hernández
University of A Coruña, Spain
Member of WIT Board of Directors

Jeff De Hosson
University of Groningen, The Netherlands

Derek O. Northwood
University of Windsor, Canada

INTERNATIONAL SCIENTIFIC ADVISORY COMMITTEE

Amelia Almeida
Djamila Benyerou
Cristina Bignardi
Ieda Caminha
Franco Concli
Marcello Conte
Patrick De Wilde
Dawood Desai
Howie Fang
Matthieu Horgnies

Frank Huberth
Zulfiqar Ahmad Khan
Ants Koel
Alireza Maheri
Vaclav Ocelik
Jose Ortiz-Lozano
Toomas Rang
Adil Saeed
Hitoshi Takagi
Kenichi Takemura

ORGANISED BY

Wessex Institute, UK
University of Groningen, The Netherlands
University of Windsor, Canada

SPONSORED BY

WIT Transactions on Engineering Sciences
*International Journal of Computational Methods and
Experimental Measurements*

WIT Transactions

Wessex Institute
Ashurst Lodge, Ashurst
Southampton SO40 7AA, UK

Senior Editors

H. Al-Kayiem
Universiti Teknologi PETRONAS, Malaysia

G. M. Carlomagno
University of Naples Federico II, Italy

A. H-D. Cheng
University of Mississippi, USA

J. J. Connor
Massachusetts Institute of Technology, USA

J. Th M. De Hosson
University of Groningen, Netherlands

P. De Wilde
Vrije Universiteit Brussel, Belgium

N. A. Dumont
PUC-Rio, Brazil

A. Galiano-Garrigos
University of Alicante, Spain

F. Garzia
University of Rome "La Sapienza", Italy

M. Hadfield
University of Bournemouth, UK

S. Hernández
University of A Coruña, Spain

J. T. Katsikadelis
National Technical University of Athens, Greece

J. W. S. Longhurst
University of the West of England, UK

E. Magaril
Ural Federal University, Russia

S. Mambretti
Politecnico di Milano, Italy

W. J. Mansur
Federal University of Rio de Janeiro, Brazil

J. L. Miralles i Garcia
Universitat Politècnica de València, Spain

G. Passerini
Università Politecnica delle Marche, Italy

F. D. Pineda
Complutense University, Spain

D. Poljak
University of Split, Croatia

F. Polonara
Università Politecnia delle Marche, Italy

D. Proverbs
Birmingham City University, UK

T. Rang
Tallinn Technical University, Estonia

G. Rzevski
The Open University, UK

P. Skerget
University of Maribor, Slovenia

B. Sundén
Lund University, Sweden

Y. Villacampa Esteve
Universidad de Alicante, Spain

P. Vorobieff
University of New Mexico, USA

S. S. Zubir
Universiti Teknologi Mara, Malaysia

Editorial Board

B. Abersek University of Maribor, Slovenia

Y. N. Abousleiman University of Oklahoma, USA

G. Alfaro Degan Università Roma Tre, Italy

K. S. Al Jabri Sultan Qaboos University, Oman

D. Almorza Gomar University of Cadiz, Spain

J. A. C. Ambrosio IDMEC, Portugal

A. M. Amer Cairo University, Egypt

S. A. Anagnostopoulos University of Patras, Greece

E. Angelino A.R.P.A. Lombardia, Italy

H. Antes Technische Universitat Braunschweig, Germany

M. A. Atherton South Bank University, UK

A. G. Atkins University of Reading, UK

D. Aubry Ecole Centrale de Paris, France

H. Azegami Toyohashi University of Technology, Japan

J. M. Baldasano Universitat Politecnica de Catalunya, Spain

J. Barnes University of the West of England, UK

J. G. Bartzis Institute of Nuclear Technology, Greece

S. Basbas Aristotle University of Thessaloniki, Greece

A. Bejan Duke University, USA

M. P. Bekakos Democritus University of Thrace, Greece

G. Belingardi Politecnico di Torino, Italy

R. Belmans Katholieke Universiteit Leuven, Belgium

D. E. Beskos University of Patras, Greece

S. K. Bhattacharyya Indian Institute of Technology, India

H. Bjornlund University of South Australia, Australia

E. Blums Latvian Academy of Sciences, Latvia

J. Boarder Cartref Consulting Systems, UK

B. Bobee Institut National de la Recherche Scientifique, Canada

H. Boileau ESIGEC, France

M. Bonnet Ecole Polytechnique, France

C. A. Borrego University of Aveiro, Portugal

A. R. Bretones University of Granada, Spain

F-G. Buchholz Universitat Gesanthochschule Paderborn, Germany

F. Butera Politecnico di Milano, Italy

W. Cantwell Liverpool University, UK

C. Capilla Universidad Politecnica de Valencia, Spain

D. J. Cartwright Bucknell University, USA

P. G. Carydis National Technical University of Athens, Greece

J. J. Casares Long Universidad de Santiago de Compostela, Spain

A. Chakrabarti Indian Institute of Science, India

F. Chejne National University, Colombia

J-T. Chen National Taiwan Ocean University, Taiwan

J. Chilton University of Lincoln, UK

C-L. Chiu University of Pittsburgh, USA

H. Choi Kangnung National University, Korea

A. Cieslak Technical University of Lodz, Poland

C. Clark Wessex Institute, UK

S. Clement Transport System Centre, Australia

M. C. Constantinou State University of New York at Buffalo, USA

M. da C Cunha University of Coimbra, Portugal

W. Czyczula Krakow University of Technology, Poland

L. D'Acierno Federico II University of Naples, Italy

M. Davis Temple University, USA

A. B. de Almeida Instituto Superior Tecnico, Portugal

L. De Biase University of Milan, Italy

R. de Borst Delft University of Technology, Netherlands

G. De Mey University of Ghent, Belgium

A. De Naeyer Universiteit Ghent, Belgium

N. De Temmerman Vrijie Universiteit Brussel, Belgium

D. De Wrachien State University of Milan, Italy

L. Debnath University of Texas-Pan American, USA

G. Degrande Katholieke Universiteit Leuven, Belgium

S. del Giudice University of Udine, Italy

M. Domaszewski Universite de Technologie de Belfort-Montbeliard, France

K. Dorow Pacific Northwest National Laboratory, USA

W. Dover University College London, UK

C. Dowlen South Bank University, UK

J. P. du Plessis University of Stellenbosch, South Africa

R. Duffell University of Hertfordshire, UK

A. Ebel University of Cologne, Germany

V. Echarri University of Alicante, Spain

K. M. Elawadly Alexandria University, Egypt

D. Elms University of Canterbury, New Zealand

M. E. M El-Sayed Kettering University, USA

D. M. Elsom Oxford Brookes University, UK

F. Erdogan Lehigh University, USA

J. W. Everett Rowan University, USA

M. Faghri University of Rhode Island, USA

R. A. Falconer Cardiff University, UK

M. N. Fardis University of Patras, Greece

A. Fayvisovich Admiral Ushakov Maritime State University, Russia

H. J. S. Fernando Arizona State University, USA

W. F. Florez-Escobar Universidad Pontifica Bolivariana, South America

E. M. M. Fonseca Instituto Politécnico do Porto, Instituto Superior de Engenharia do Porto, Portugal

D. M. Fraser University of Cape Town, South Africa

G. Gambolati Universita di Padova, Italy

C. J. Gantes National Technical University of Athens, Greece

L. Gaul Universitat Stuttgart, Germany

N. Georgantzis Universitat Jaume I, Spain

L. M. C. Godinho University of Coimbra, Portugal

F. Gomez Universidad Politecnica de Valencia, Spain

A. Gonzales Aviles University of Alicante, Spain

D. Goulias University of Maryland, USA

K. G. Goulias Pennsylvania State University, USA

W. E. Grant Texas A & M University, USA

S. Grilli University of Rhode Island, USA

R. H. J. Grimshaw Loughborough University, UK

D. Gross Technische Hochschule Darmstadt, Germany

R. Grundmann Technische Universitat Dresden, Germany

O. T. Gudmestad University of Stavanger, Norway

R. C. Gupta National University of Singapore, Singapore

J. M. Hale University of Newcastle, UK

K. Hameyer Katholieke Universiteit Leuven, Belgium

C. Hanke Danish Technical University, Denmark

Y. Hayashi Nagoya University, Japan

L. Haydock Newage International Limited, UK

A. H. Hendrickx Free University of Brussels, Belgium

C. Herman John Hopkins University, USA

I. Hideaki Nagoya University, Japan

W. F. Huebner Southwest Research Institute, USA

M. Y. Hussaini Florida State University, USA

W. Hutchinson Edith Cowan University, Australia

T. H. Hyde University of Nottingham, UK

M. Iguchi Science University of Tokyo, Japan

L. Int Panis VITO Expertisecentrum IMS, Belgium

N. Ishikawa National Defence Academy, Japan

H. Itoh University of Nagoya, Japan

W. Jager Technical University of Dresden, Germany

Y. Jaluria Rutgers University, USA

D. R. H. Jones University of Cambridge, UK

N. Jones University of Liverpool, UK

D. Kaliampakos National Technical University of Athens, Greece

D. L. Karabalis University of Patras, Greece

A. Karageorghis University of Cyprus

T. Katayama Doshisha University, Japan

K. L. Katsifarakis Aristotle University of Thessaloniki, Greece

E. Kausel Massachusetts Institute of Technology, USA

H. Kawashima The University of Tokyo, Japan

B. A. Kazimee Washington State University, USA

F. Khoshnaw Koya University, Iraq

S. Kim University of Wisconsin-Madison, USA

D. Kirkland Nicholas Grimshaw & Partners Ltd, UK

E. Kita Nagoya University, Japan

A. S. Kobayashi University of Washington, USA

D. Koga Saga University, Japan

S. Kotake University of Tokyo, Japan

A. N. Kounadis National Technical University of Athens, Greece

W. B. Kratzig Ruhr Universitat Bochum, Germany

T. Krauthammer Penn State University, USA

R. Laing Robert Gordon University, UK

M. Langseth Norwegian University of Science and Technology, Norway

B. S. Larsen Technical University of Denmark, Denmark

F. Lattarulo Politecnico di Bari, Italy

A. Lebedev Moscow State University, Russia

D. Lesnic University of Leeds, UK

D. Lewis Mississippi State University, USA

K-C. Lin University of New Brunswick, Canada

A. A. Liolios Democritus University of Thrace, Greece

D. Lippiello Università degli Studi Roma Tre, Italy

S. Lomov Katholieke Universiteit Leuven, Belgium

J. E. Luco University of California at San Diego, USA

L. Lundqvist Division of Transport and Location Analysis, Sweden

T. Lyons Murdoch University, Australia

L. Mahdjoubi University of the West of England, UK

Y-W. Mai University of Sydney, Australia

M. Majowiecki University of Bologna, Italy

G. Manara University of Pisa, Italy

B. N. Mandal Indian Statistical Institute, India

Ü. Mander University of Tartu, Estonia

H. A. Mang Technische Universitat Wien, Austria

G. D. Manolis Aristotle University of Thessaloniki, Greece

N. Marchettini University of Siena, Italy

J. D. M. Marsh Griffith University, Australia

J. F. Martin-Duque Universidad Complutense, Spain

T. Matsui Nagoya University, Japan

G. Mattrisch DaimlerChrysler AG, Germany

F. M. Mazzolani University of Naples "Federico II", Italy

K. McManis University of New Orleans, USA

A. C. Mendes Universidade de Beira Interior, Portugal

J. Mera Polytechnic University of Madrid, Spain

J. Mikielewicz Polish Academy of Sciences, Poland

R. A. W. Mines University of Liverpool, UK

C. A. Mitchell University of Sydney, Australia

K. Miura Kajima Corporation, Japan

A. Miyamoto Yamaguchi University, Japan

T. Miyoshi Kobe University, Japan

G. Molinari University of Genoa, Italy

F. Mondragon Antioquin University, Colombia

T. B. Moodie University of Alberta, Canada

D. B. Murray Trinity College Dublin, Ireland

M. B. Neace Mercer University, USA

D. Necsulescu University of Ottawa, Canada

B. Ning Beijing Jiatong University, China

S-I. Nishida Saga University, Japan

H. Nisitani Kyushu Sangyo University, Japan

B. Notaros University of Massachusetts, USA

P. O'Donoghue University College Dublin, Ireland

R. O. O'Neill Oak Ridge National Laboratory, USA

M. Ohkusu Kyushu University, Japan

G. Oliveto Universitá di Catania, Italy

R. Olsen Camp Dresser & McKee Inc., USA

E. Oñate Universitat Politecnica de Catalunya, Spain

K. Onishi Ibaraki University, Japan

P. H. Oosthuizen Queens University, Canada

E. Outa Waseda University, Japan

O. Ozcevik Istanbul Technical University, Turkey

A. S. Papageorgiou Rensselaer Polytechnic Institute, USA

J. Park Seoul National University, Korea

F. Patania Universitá di Catania, Italy

B. C. Patten University of Georgia, USA

G. Pelosi University of Florence, Italy

G. G. Penelis Aristotle University of Thessaloniki, Greece

W. Perrie Bedford Institute of Oceanography, Canada

M. F. Platzer Naval Postgraduate School, USA

D. Prandle Proudman Oceanographic Laboratory, UK

R. Pulselli University of Siena, Italy

I. S. Putra Institute of Technology Bandung, Indonesia

Y. A. Pykh Russian Academy of Sciences, Russia

A. Rabasa University Miguel Hernandez, Spain

F. Rachidi EMC Group, Switzerland

K. R. Rajagopal Texas A & M University, USA

J. Ravnik University of Maribor, Slovenia

A. M. Reinhorn State University of New York at Buffalo, USA

G. Reniers Universiteit Antwerpen, Belgium

A. D. Rey McGill University, Canada

D. N. Riahi University of Illinois at Urbana-Champaign, USA

B. Ribas Spanish National Centre for Environmental Health, Spain

K. Richter Graz University of Technology, Austria

S. Rinaldi Politecnico di Milano, Italy

F. Robuste Universitat Politecnica de Catalunya, Spain

A. C. Rodrigues Universidade Nova de Lisboa, Portugal

G. R. Rodríguez Universidad de Las Palmas de Gran Canaria, Spain

C. W. Roeder University of Washington, USA

J. M. Roesset Texas A & M University, USA

W. Roetzel Universitaet der Bundeswehr Hamburg, Germany

V. Roje University of Split, Croatia

R. Rosset Laboratoire d'Aerologie, France

J. L. Rubio Centro de Investigaciones sobre Desertificacion, Spain

T. J. Rudolphi Iowa State University, USA

S. Russenchuck Magnet Group, Switzerland

H. Ryssel Fraunhofer Institut Integrierte Schaltungen, Germany

S. G. Saad American University in Cairo, Egypt

M. Saiidi University of Nevada-Reno, USA

R. San Jose Technical University of Madrid, Spain

F. J. Sanchez-Sesma Instituto Mexicano del Petroleo, Mexico

B. Sarler Nova Gorica Polytechnic, Slovenia

S. A. Savidis Technische Universitat Berlin, Germany

A. Savini Universita de Pavia, Italy

G. Schleyer University of Liverpool, UK

R. Schmidt RWTH Aachen, Germany

B. Scholtes Universitaet of Kassel, Germany

A. P. S. Selvadurai McGill University, Canada

J. J. Sendra University of Seville, Spain

S. M. Şener Istanbul Technical University, Turkey

J. J. Sharp Memorial University of Newfoundland, Canada

Q. Shen Massachusetts Institute of Technology, USA

G. C. Sih Lehigh University, USA

L. C. Simoes University of Coimbra, Portugal

A. C. Singhal Arizona State University, USA

J. Sladek Slovak Academy of Sciences, Slovakia

V Sladek Slovak Academy of Sciences, Slovakia

A. C. M. Sousa University of New Brunswick, Canada

H. Sozer Illinois Institute of Technology, USA

P. D. Spanos Rice University, USA

T. Speck Albert-Ludwigs-Universitaet Freiburg, Germany

C. C. Spyrakos National Technical University of Athens, Greece

G. E. Swaters University of Alberta, Canada

S. Syngellakis Wessex Institute, UK

J. Szmyd University of Mining and Metallurgy, Poland

H. Takemiya Okayama University, Japan

I. Takewaki Kyoto University, Japan

C-L. Tan Carleton University, Canada

E. Taniguchi Kyoto University, Japan

S. Tanimura Aichi University of Technology, Japan

J. L. Tassoulas University of Texas at Austin, USA

M. A. P. Taylor University of South Australia, Australia

A. Terranova Politecnico di Milano, Italy

T. Tirabassi National Research Council, Italy

S. Tkachenko Otto-von-Guericke-University, Germany

N. Tomii Chiba Institute of Technology, Japan

T. Tran-Cong University of Southern Queensland, Australia

R. Tremblay Ecole Polytechnique, Canada

I. Tsukrov University of New Hampshire, USA

R. Turra CINECA Interuniversity Computing Centre, Italy

S. G. Tushinski Moscow State University, Russia

R. van der Heijden Radboud University, Netherlands

R. van Duin Delft University of Technology, Netherlands

P. Vas University of Aberdeen, UK

R. Verhoeven Ghent University, Belgium

A. Viguri Universitat Jaume I, Spain

S. P. Walker Imperial College, UK

G. Walters University of Exeter, UK

B. Weiss University of Vienna, Austria

T. W. Wu University of Kentucky, USA

S. Yanniotis Agricultural University of Athens, Greece

A. Yeh University of Hong Kong, China

B. W. Yeigh University of Washington, USA

K. Yoshizato Hiroshima University, Japan

T. X. Yu Hong Kong University of Science & Technology, Hong Kong

M. Zador Technical University of Budapest, Hungary

R. Zainal Abidin Infrastructure University Kuala Lumpur, Malaysia

K. Zakrzewski Politechnika Lodzka, Poland

M. Zamir University of Western Ontario, Canada

G. Zappalà National Research Council, Italy

R. Zarnic University of Ljubljana, Slovenia

Materials and Contact Characterisation X

Editors

Santiago Hernández
University of A Coruña, Spain
Member of WIT Board of Directors

Jeff De Hosson
University of Groningen, The Netherlands

Derek O. Northwood
University of Windsor, Canada

WITPRESS Southampton, Boston

Editors:

Santiago Hernández
University of A Coruña, Spain
Member of WIT Board of Directors

Jeff De Hosson
University of Groningen, The Netherlands

Derek O. Northwood
University of Windsor, Canada

Published by

WIT Press
Ashurst Lodge, Ashurst, Southampton, SO40 7AA, UK
Tel: 44 (0) 238 029 3223; Fax: 44 (0) 238 029 2853
E-Mail: witpress@witpress.com
http://www.witpress.com

For USA, Canada and Mexico

Computational Mechanics International Inc
25 Bridge Street, Billerica, MA 01821, USA
Tel: 978 667 5841; Fax: 978 667 7582
E-Mail: infousa@witpress.com
http://www.witpress.com

British Library Cataloguing-in-Publication Data

A Catalogue record for this book is available
from the British Library

ISBN: 978-1-78466-437-4
eISBN: 978-1-78466-438-1
ISSN: 1746-4471 (print)
ISSN: 1743-3533 (on-line)

The texts of the papers in this volume were set individually by the authors or under their supervision. Only minor corrections to the text may have been carried out by the publisher.

No responsibility is assumed by the Publisher, the Editors and Authors for any injury and/or damage to persons or property as a matter of products liability, negligence or otherwise, or from any use or operation of any methods, products, instructions or ideas contained in the material herein.

© WIT Press 2021

Open Access: All of the papers published in this journal are freely available, without charge, for users to read, download, copy, distribute, print, search, link to the full text, or use for any other lawful purpose, without asking prior permission from the publisher or the author as long as the author/copyright holder is attributed. This is in accordance with the BOAI definition of open access.

Creative Commons content: The CC BY 4.0 licence allows users to copy, distribute and transmit an article, and adapt the article as long as the author is attributed. The CC BY licence permits commercial and non-commercial reuse.

Preface

This volume contains some of the contributions to the 10th International Conference on Computational Methods and Experiments in Materials Characterisation (MC), organised by the Wessex Institute in collaboration with the University of Groningen and the University of Windsor. The initial venue of the conference was the city of Madrid (Spain) but the situation created by the COVID-19 pandemic made necessary to transform the in person event into an online forum.

This series of conferences began in Santa Fe, New Mexico in 2003, followed by conferences in Portland, Maine (2004); Bologna (2007); the New Forest, home of the Wessex Institute of Technology (2009); Kos, Greece (2011); La Certosa di Pontignano of the University of Siena (2013), Valencia (2015),Tallinn (2017) and Lisbon (2019).

The contents of this Volume reflect the rapid advances that have taken place in materials science and engineering, prompted by the demand for high quality performance materials by industry. Some contributions present research carried out in mechanical testing and characterizations of several classes of materials including high performance steel, biomaterials, biofillers, cotton, polymers or rammed earth. Validation procedures using numerical methods as finite element models in linear and non-linear theory are also described.

Studies on recycled and reclaimed materials are included in this book relating reuse of truck tyres, roof waste, ash from MSW incineration or polyethylene. Emerging and green materials are also the subject of some papers showing advances in applications of nanocellulose, biomass, smart textiles and bio-algae plastic.

The effectiveness of various surface treatments to enhance materials behaviour was discussed and techniques for designing performance base material were shown. The wide range of topics includes interaction between disciplines, which is sometimes essential to achieving a proper understanding of material behaviour.

The Meeting was sponsored by WIT Transactions on Engineering Sciences and the International Journal of Computational Methods and Experimental Measurements.

The conference addressed these problems continuing to expand on the development of previous meetings in the series. Papers presented at MC are a relevant addition to the state of the art in this field.

All papers published in this Volume, as well as those of the previous conferences, have been published in paper as well as digital format and are being widely distributed throughout the world. They are permanently archived at www.witpress.com/elibrary, where they are easily available to the international community. As with other papers presented at Wessex Institute conferences, they are part of the WIT Transactions, a collection that appears in notable reviews, publications and databases.

The Editors are grateful to the members of the International Scientific Advisory Committee and the reviewers that made an outstanding job selecting the papers in this book, as well as to all authors for their collaboration.

The Editors
Ashurst Lodge, 2021

Contents

Section 3: Emerging and green materials

SECTION 1
MECHANICAL TESTING
AND CHARACTERISATION

MOUTH-LIKE CRACKING IN A HIGH-STRENGTH MULTIPHASE STEEL AND ITS RELATIONSHIP TO FRACTURE TOUGHNESS

XUAN WANG[1], JICHENG GAO[1], YUN HANG[1], YI ZHENG[1], DEREK O. NORTHWOOD[2] & CHENG LIU[1]
[1]School of Mechanical Engineering, Yangzhou University, People's Republic of China
[2]Mechanical, Auto and Materials Engineering, University of Windsor, Canada

ABSTRACT

The effect of microstructure on crack growth behaviour in steels has always been a subject of considerable research interest. Based on a quenching and partitioning process (Q&P), and the transformation of a nano-scaled bainite in advanced high-strength steels, a novel quenching-partitioning-austempering process (Q-P-A) has been developed for manufacturing a multiphase microstructure in a medium carbon steel (55Mn2SiCr). The processing sequence consists of the following steps: austenitizing at 900°C for 0.5 h; controlled quenching and cooling to 200°C, i.e. slightly below Ms (the start temperature for martensite transformation) for 5 s; austempering at 170°C for 5 min; up-heating to 250°C for 120 min; final air cooling to room temperature. An ultimate tensile strength (UTS) above 2 GPa, as well as an acceptable elongation of 3%, is obtained due to a multiphase formation comprising prior martensite (PM), bainitic ferrite (BF), retained austenite (RA) and nano-scaled structure ((BF + RA$_{(+C)}$)nano). Mouth-like cracks are observed on the fracture surface and the crack arrest behavior is investigated. When a microstructural cluster with (BF + RA$_{(+C)}$)nano fully covered PM is formed, a mouth-like crack can be formed and a superior crack resistance can be obtained. The crack initiates from the PM boundary and propagates along the interface between the PM and (BF + RA$_{(+C)}$)nano over a distance of a few millimeters and before being arrested in the (BF + RA$_{(+C)}$)nano. This behaviour is mainly attributed to the uniform distribution of film RA and needle BF with nano-level spacing in the (BF + RA)nano. The stress concentration energy at the crack tip can be absorbed by the martensitic transformation of the film RA. The results are important when designing a multiphase microstructure for a commercial high-strength steel.

Keywords: high tensile strength, mouth-like crack, quenching-partitioning-austempering process, nano-scaled structure, crack initiation and propagation.

1 INTRODUCTION

Advanced high-strength steels (AHSS) have been widely investigated due to their excellent strength, good ductility and high strength to weight ratio as majority of engineering designs require structural materials to have high strength and fracture toughness while minimizing weight. The latest generation of AHSS, also called the third generation of AHSS, produced by quenching and partitioning (Q&P) process, can provide a better strength and ductility combination compared to the first generation, while with a lower cost than the second generation [1]. The Q&P process was proposed by Speer et al. [2] in 2003, and consisted of an interrupted quenching process between the martensite-start temperature (Ms) and the martensite-finish temperature (Mf) after full austenitization or intercritical annealing. Accordingly, new AHSS have been developed based on the phase transition mechanism in the Q&P process for improved properties [3].

Nowadays, among the various competitive strategies to obtain AHSS, a low temperature isothermal treatment for producing a super bainitic steel has received considerable attention [4]. Such steels have excellent mechanical properties, as well as an affordable price. This is mainly due to the nano-scaled microstructures comprising bainitic ferrite (BF) sheaves and carbon-enriched retained austenite (RA) trapped between BF laths. However, the long isothermal treatment time, over 60 days, would be totally unacceptable for commercial manufacture. Navarro-López et al. [5] have pointed out that prior martensite (PM), which is formed during quenching step in the process of Q&P treatment, could accelerate the bainite transformation kinetics by creating potential nucleation sites at the martensite-austenite interfaces, in addition to the austenite boundaries. Generally, a multiphase microstructure comprising PM, bainitic ferrite (BF) plus retained austenite (RA) is considered as an attractive microstructure for the third generation of advanced high-strength steels [6]–[9].

The influence of microstructure on crack propagation behavior in the high-performance steels has been an important research field [10]–[12]. Recently, Zhao et al. [13] investigated the fatigue failure mechanism of a multiphase steel. They found that the crack initiation site within a multiphase structure was the dominant factor influencing the high fatigue strength. Yajima et al. [14] in their research on the crack nucleation mechanism of austempered ductile iron during tensile deformation, reported that the cracks could benefit to the stress-induced martensitic transformation. Traditionally, it is believed that cracks are harmful to the properties of steel, especially for AHSS which requires not only high tensile strength but also excellent ductility [15]–[17].

In this paper, the cracks morphologies are studied in a medium carbon steel (55Mn2SiCr) that was heat treated using a quenching-partitioning-austempering (Q-P-A) process, that was developed in-house. The relationship between the initiation and propagation of mouth-like cracks, the multiphase microstructure and the mechanical properties is discussed.

2 EXPERIMENTAL PROCEDURES

The investigated material was a 6 mm-thick, hot-rolled and normalized plate. The composition is given in Table 1. Ms point of the steel calculated using the MUCG83 thermodynamic model, is 207°C [18]. The shape and size of tensile sample is shown in Fig. 1, which is designed according to ASTM standard E8M [19]. The scheme of the applied heat treatments is presented in Fig. 2. The processing sequence consists of the following steps: austenitizing at 900°C for 0.5 h; controlled quenching and cooling to 200°C, i.e. slightly below Ms (the start temperature for martensite transformation) for 5 s; quenching at 170°C for 5 min; up-heating to 250°C for 30, 60, 120 and 240 min respectively; final air cooling to room temperature.

A surface layer of 0.35 mm-thickness for tensile samples was removed after heat treatment and the tensile test with a 100 kN load at a crosshead speed of 0.05 mm·min^{-1} was conducted on a DNS100 universal testing machine. A HR-150A Rockwell hardness tester was used for hardness measurement with a 150 kg load and 120° diamond cone indenter. Microhardness tests were made by a Sinowon HV-1000A machine with a diamond square cone indenter and a load of 0.98 N.

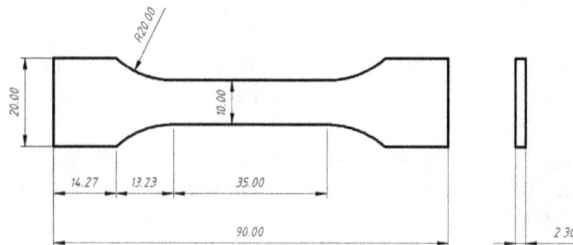

Figure 1: Geometry and dimensions (unit: mm) of tensile specimen.

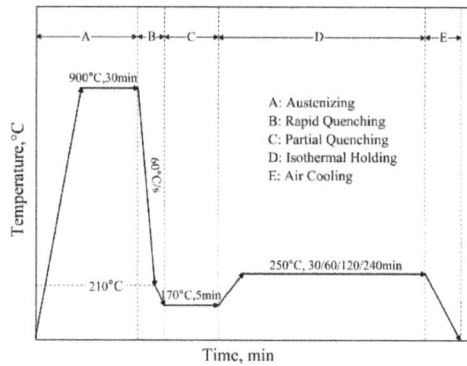

Figure 2: Schematic for heat treatment process.

Table 1: Chemical composition of sample (wt. %).

C	Si	Mn	Cr	V	Mo	P	S	Fe
0.57	1.50	1.83	0.83	0.03	0.01	0.01	0.007	Bal.

After being ground, polished and etched in 2% Nital solution, the microstructural characteristics of the samples were examined using optical microscopy (OM, LEICA GQ-300). The fracture surfaces of all samples were observed on a XL30-ESEM scanning electron microscope (SEM). Quantitative analysis of the microstructures were made using Image-Pro Plus 6.0 software.

3 RESULTS AND DISCUSSION

3.1 Microstructure

OM micrographs of the as-received samples after hot-rolling and normalizing are presented in Fig. 3. The microstructure is mainly composed of pearlite and ferrite. The dark gray area in Fig. 3(a) is pearlite and the white area is ferrite. From the grey area in Fig. 3(b), the lamella structure of pearlite can be clearly observed.

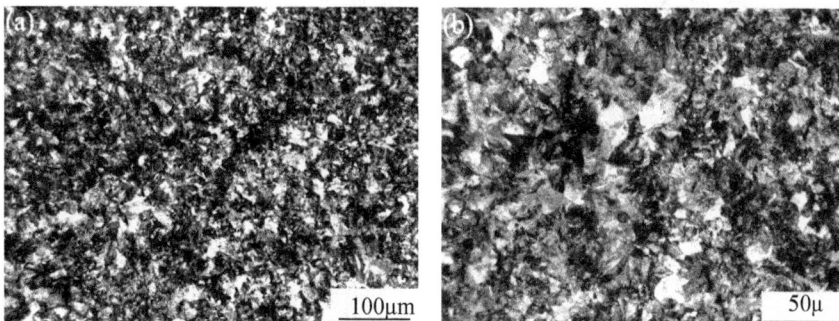

Figure 3:   OM micrographs of as-received steel. (a) Low magnification; and (b) High magnification.

Fig. 4 shows that the OM micrographs of the samples after austempering at 250°C for various times. Typical microstructures consist of prior martensite (PM), bainitic ferrite (BF) and retained austenite (RA). With increasing tempering time, the content of PM does not appear to change, and the amount of BF increases. RA exists in the form of blocks or films. With increasing the tempering time, the content of blocky RA decreases, but the film RA increases. Corresponding SEM micrographs are presented in Fig. 5. PM is dark, which has obvious boundaries with the surrounding phases, and its internal substructure can be observed. RA appears as a grey phase with an overall content tending to decrease with time. Fig. 6 shows the detailed microstructure of the sample after austempering at 250°C for 60 min. Carbide precipitation is evident in the PM.

3.2 Mechanical properties

Fig. 7 summarises the hardness and microhardness results for samples austempered for 30, 60, 120 and 240 min at 250°C. The hardness (HRC) shows a small decrease, 60 to 55 HRC, with increasing austempering time. The microhardness measurements allowed us to probe the hardness of the constituent phases (RA, BF, PM). The RA showed the largest hardness decrease of the constituent phases, decreasing from 800 HV for 30 min austempering, to 600–650 HV after 60 min austempering. The PM hardness decreased from 770 HV after 30 min austempering to about 700 HV after 120 min austempering. The BF hardness remained approximately constant at a level of 670 to 700 HV.

Figs 8 and 9 summarise the mechanical properties (YS, TS, % elongation). The YS and TS both increase with austempering time up to an austempering time of 120 min. Austempering for 240 min produces a substantive reduction in both YS and TS. The % elongation initially increases with austempering time from 30 min to 60 min, but then

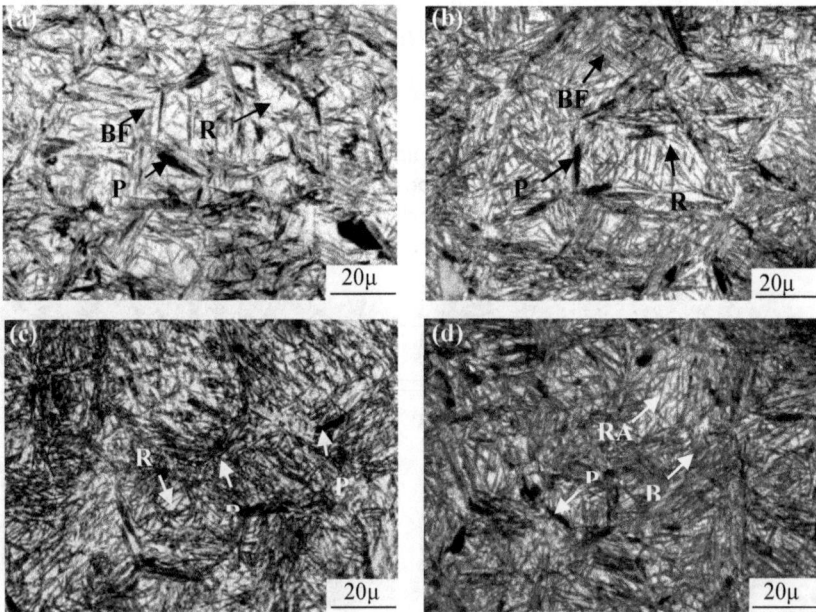

Figure 4:   OM micrographs of samples after austempering at 250°C for (a) 30 min; (b) 60 min; (c) 120 min; and (d) 240 min.

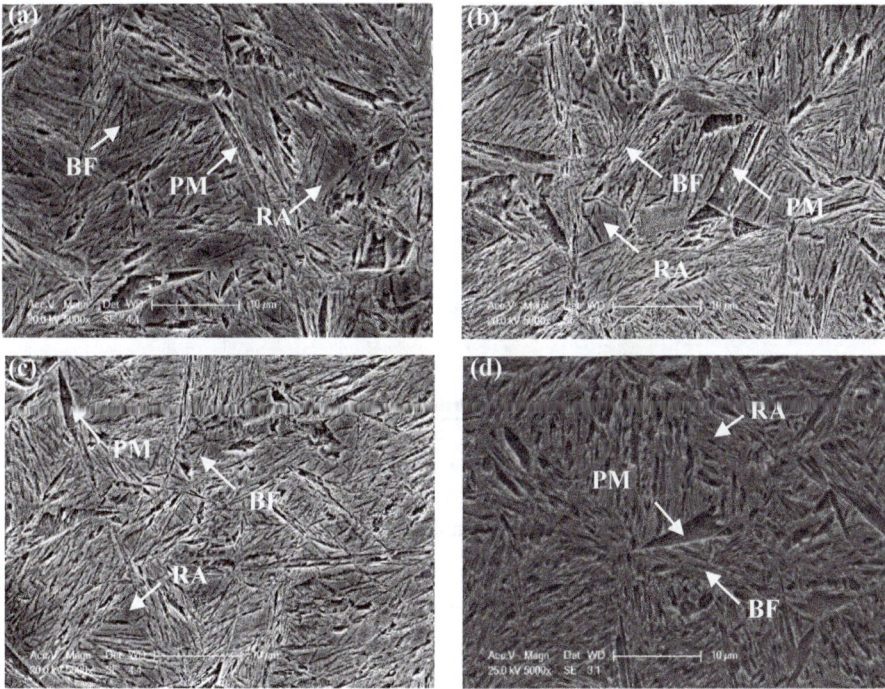

Figure 5:   SEM micrographs of samples after austempering at 250°C for (a) 30 min; (b) 60 min; (c) 120 min and (d) 240 min.

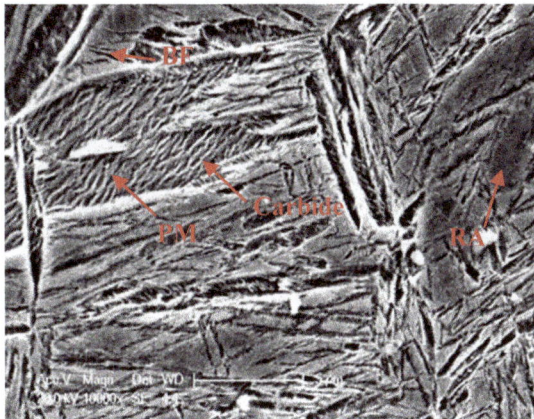

Figure 6: High magnification SEM micrograph of sample at 250°C for 60 min.

decreases with a further increase in austempering time. The optimum mechanical properties are produced by austempering at 250°C for 120 min: YS 1,684 MPa; TS 2,030 MPa; 3% elongation.

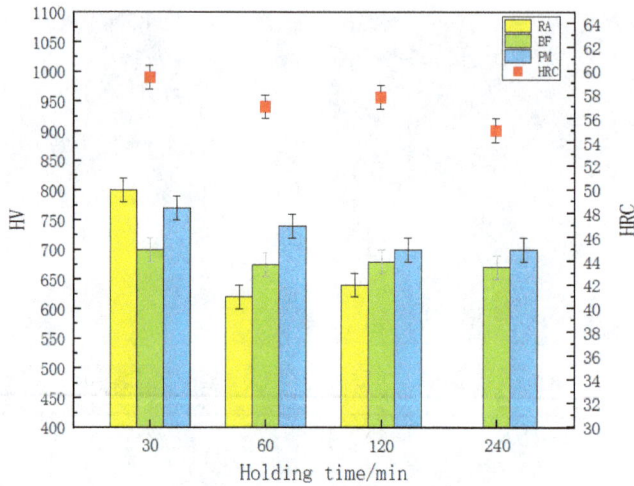

Figure 7: Hardness of samples after austempering at 250°C for different times.

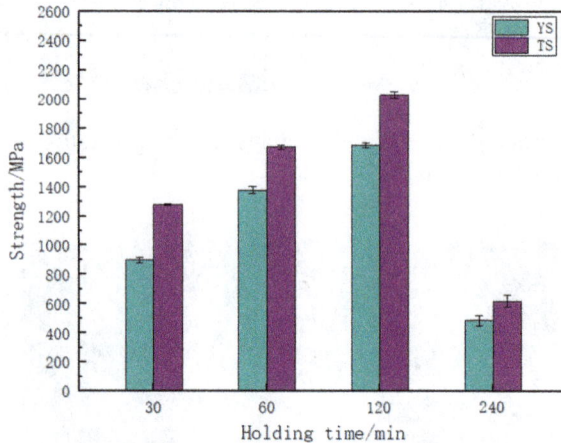

Figure 8: Mechanical properties of samples after austempering at 250°C for different times.

3.3 Mouth-like cracking and mechanisms of failure

The fracture surfaces of samples austempered for various times are shown in Figs 10–13. In the sample austempered at 250°C for 30 min, a ductile and brittle fracture mode is observed (see Fig. 10). The red circle indicated in Fig. 10(a) is in the sheared lip zone, which is mainly composed of small and uniform dimples. Nearby is a small crack, shown by the green rectangle area in Fig. 10(a). This crack has flat upper and lower surfaces, and looks like a mouth with two open corners (crack tips). It is about 30 μm long and 3.0 μm wide. Mixed brittle and ductile fracture of the upper face and a cleavage characteristic in the lower face of the mouth-like crack were observed. A similar crack was found in the sample austempered

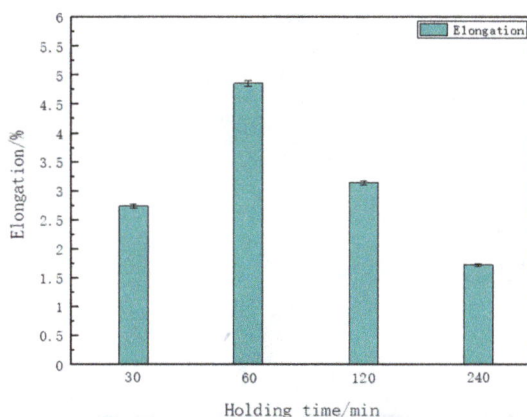

Figure 9: Elongation of samples after austempering at 250°C for different times.

for 60 min, Fig. 11. It had a length of about 20 μm and a width of 4.0 μm. At a high magnification, Fig. 11(b) and (c), it was evident that the sample austempered for 60 min had more dimples on both fracture surfaces than that for 30 min (Fig. 10(c)).

It can be seen from Fig. 12 that the morphology of the mouth-like crack observed in austempered sample for 120 min at 250°C is different from those shown in Figs 10 and 11 in two respects. First, the upper or lower surface of mouth-like crack is full of small, uniform dimples, indicating a ductile failure mode. Second, the two mouth corners of the crack are closed, forming a crack closure loop. The crack length was about 75 μm and the crack width was 15 μm. Fig. 13 illustrates the morphology of a large crack in the sample austempered for

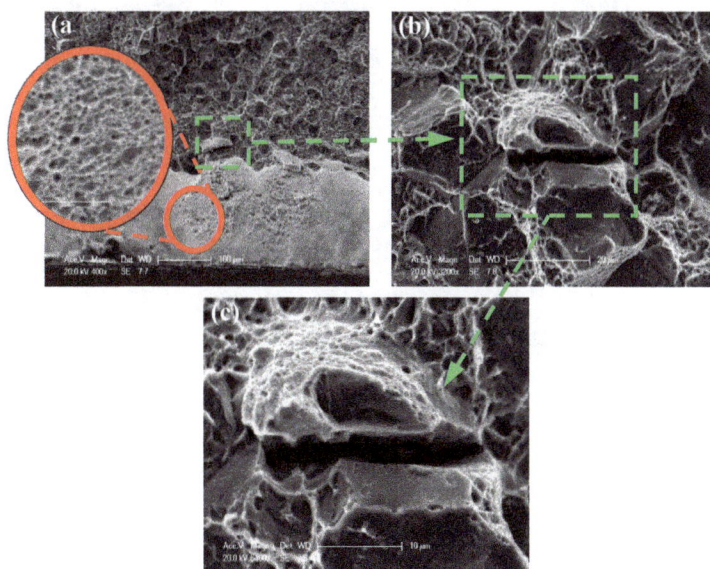

Figure 10: SEM micrographs of tensile fracture of sample austempered at 250°C for 30 min. (a) 400×; (b) 3,200×; and (c) 6,400×.

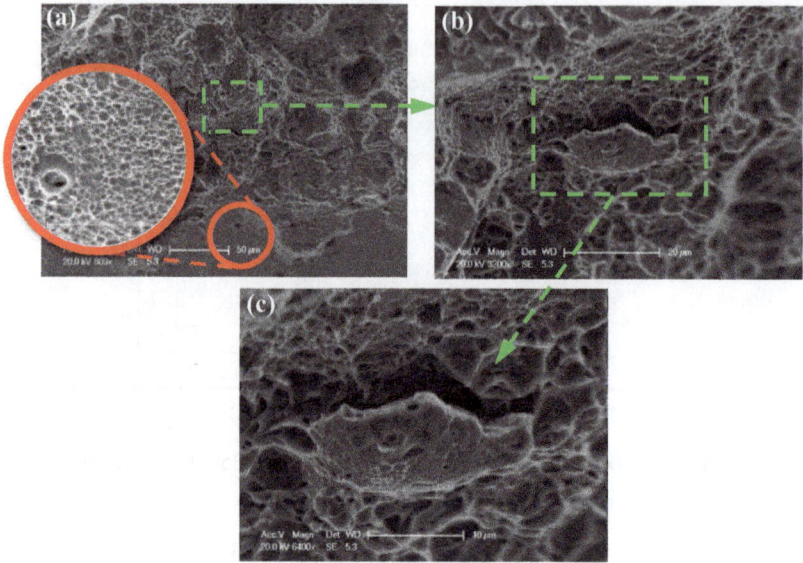

Figure 11: SEM micrographs of tensile fracture of sample austempered at 250°C for 60 min. (a) 800×; (b) 3,200×; and (c) 6,400×.

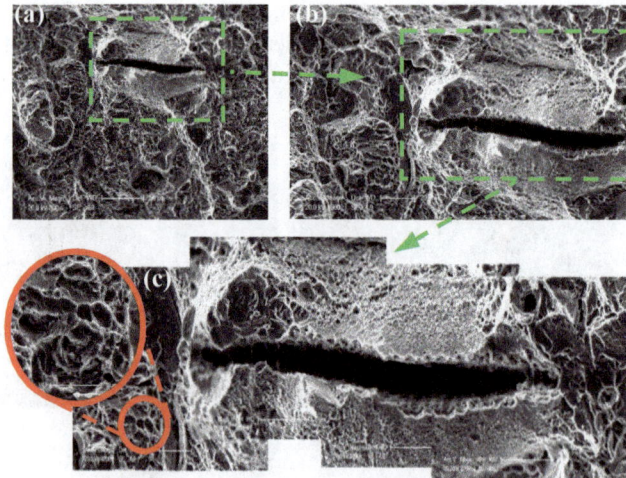

Figure 12: SEM micrographs of tensile fracture of sample at 250°C for 120 min. (a) 800×; (b) 1,600×; and (c) 3,200×.

240 min at 250°C, which is non-mouth-like and consisted of a main crack (100 μm long and 20 μm wide) and a secondary crack (50 μm long and 1.0 μm wide). The fracture morphology on both sides of the main crack is quite different. The crack upper side is dominated by dimples with different scales and depths while the lower side indicates a cleavage fracture mode.

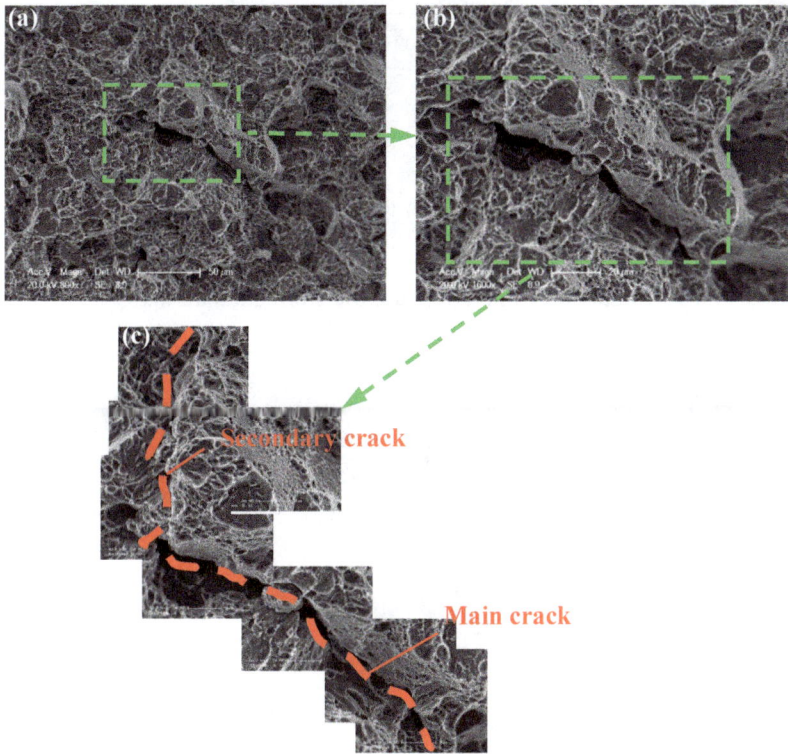

Figure 13:   SEM micrographs of tensile fracture of sample austempered at 250°C for 240 min. (a) 800×; (b) 1,600×; and (c) 6,400×.

A schematic of the phase transformation during Q-P-A process is presented in Fig. 14. The main purpose of the rapid cooling (see Fig. 2(b)) is to avoid the formation of non-martensite phases such as pearlite and ferrite. About 30% prior martensite (PM) can be produced from austenite by partial quenching to 180°C (below the Ms), based on the calculation in ref. [20] and our previous experimental results [21] (see Fig. 15(a)). When the sample is austempered at 250°C, the supersaturated carbon atoms in the PM partially precipitate within the martensitic lath and partially partition through the phase boundaries to the austenite (A), due to a high Si content of the 55Mn2SiCr alloy. With respect to the partially tempered PM, it is easy for BF to nucleate at a local carbon-depleted area along the boundary between PM and A. Because of a relatively low carbon diffusion rate at 250°C and an uneven carbon distribution on the phase interface, nano-scaled BF laths are formed, interwoven with A films. When the austempering time increases, the BF growth occurs by more carbon atoms diffusing from BF into A. Thus, a carbon concentration in A increases to become $A_{(+C)}$, and finally $RA_{(+C)}$ at room temperature. That is, a unique microstructural cluster is formed by the Q-P-A process based on the carbon diffusion characteristics of PM, which is composed of a nano structure (BF + $RA_{(+C)}$) nucleating around a PM ((BF + $RA_{(+C)}$)nano).

The formation of a mouth-like crack is mainly related to the nano-scaled microstructural cluster. Fig. 15 is a schematic diagram of mouth-like crack initiation and propagation. When

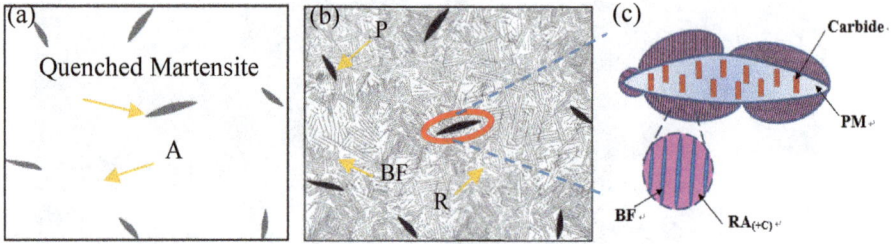

Figure 14: Schematic illustration of phase transformation during Q-P-A process.

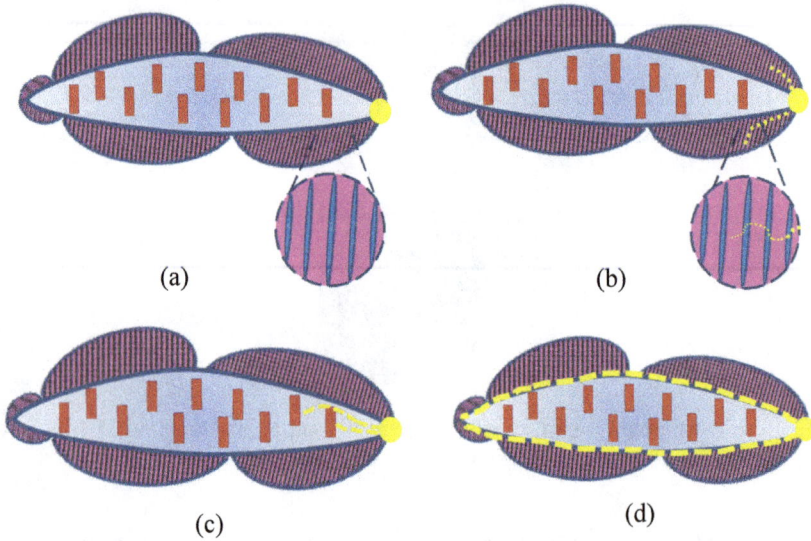

Figure 15: Schematic of mouth-like crack formation. (a) Crack initiation at tip of PM; (b) Crack propagation through (BF + RA(+C)) nanostructure surrounding the PM; (c) Crack propagation into the PM; and (d) Crack propagation along the PM-(BF + RA(+C)) nano) interface.

local plastic deformation occurs under tensile loading, the boundary between the PM and (BF + RA$_{(+C)}$) is an area where the stress concentration is high due to non-uniform deformation of the various phases, and can therefore initiate micro-cracks. Fig. 15 shows that there are three ways for crack propagation after forming on the boundary within nano-scaled structure cluster. It may propagate into the PM and its propagation is stopped or lessened when it meets the carbide phase in the martensitic lath (see Fig. 15(c)). Or it may enter into the (BF + RA$_{(+C)}$)nano around the PM. The crack propagating orientation will be changed by the influence of the fine laminated structure (see Fig. 15(b)) and finally stop. It has been reported that the multiphase nano-laminate microstructure can not only deflect the crack during propagation, but also be arrested by a metastable multiphase martensite-austenite transformation-induced plasticity (TRIP) effect [22]. The stress concentration energy at the crack tip can be absorbed by the martensitic transformation of the film RA.

Also, the micro-crack can propagate preferentially along the PM-(BF + RA$_{(+C)}$)nano interface (see Fig. 15(d)), since the BF nucleation occurs close to rather than at the interface [23]. An extremely small austenite area with an inhomogeneous carbon concentration can provide a channel for crack growth. If the PM is fully covered by the (BF + RA$_{(+C)}$)nano, the crack will be arrested with the structure cluster when the crack reaches the boundaries of (BF + RA$_{(+C)}$) nano, resulting in a mouth-like crack with closed corners. Thus, the tensile strength is increased. If not, a mouth-like crack with open corners is formed, which is attributed to a relatively lower mechanical property. This is supported by the mechanical property data in Figs 7 and 9. The poor mechanical properties of the sample austempered for 240 min are probably due to carbide precipitation from austenite and BF. Thus, the carefully designed multiphase microstructure has been changed, which is detrimental to the mechanical properties. Further investigation is required by XRD and TEM.

4 CONCLUSIONS

A multiphase high-strength steel is obtained by using a novel quenching-partitioning-austempering process (Q-P-A). The resultant microstructure consists of prior martensite (PM), bainitic ferrite (BF), retained austenite (RA) and (BF + RA$_{(+C)}$)nano. Mouth-like cracks are observed on the fracture surfaces of tensile test samples austempered at 250°C for 30, 60 and 120 min during tensile tests. The morphology of mouth-like cracks can influence the mechanical property. A mouth-like crack with closed corners, formed when the (BF + RA(+C))nano nucleation occurs fully around the PM, produces the optimum combination of mechanical properties.

REFERENCES

[1] Bleck, W., Brühl, F., Ma, Y. & Sasse, C., Materials and processes for the third-generation advanced high-strength steels. *Berg- und hüttenmännische Monatshefte*, **164**(11), pp. 466–474, 2019.

[2] Speer, J., Matlock, D.K., De Cooman, B.C. & Schroth, J.G., Carbon partitioning into austenite after martensitic transformation. *Acta Materialia*, **51**, pp. 2611–2622, 2003.

[3] Huyghe, P. et al., On the effect of Q&P processing on the stretch-flange-formability of 0.2C ultra-high strength steel sheets. *ISIJ International*, **58**(7), pp. 1341–1350, 2018.

[4] Avishan, B., Tavakolian, M. & Yazdani, S., Two-step austempering of high performance steel with nanoscale microstructure. *Materials Science & Engineering A*, **693**, pp. 178–185, 2017.

[5] Navarro-López, A., Sietsma, J., & Santofimia, M.J., Effect of prior athermal martensite on the isothermal transformation kinetics below Ms in a low-C high-Si steel. *Metallurgical and Materials Transactions A*, **47**, pp. 1028–1039, 2016.

[6] Diego-Calderón, I.D. et al., Effect of microstructure on fatigue behavior of advanced high strength steels produced by quenching and partitioning and the role of retained austenite. *Materials Science & Engineering A*, **641**, pp. 215–224, 2015.

[7] Amel-Farzad, H., Faridi, H.R., Rajabpour, F., Abolhasani, A., Kazemi, S. & Khaledzadeh, Y., Developing very hard nanostructured bainitic steel. *Materials Science and Engineering A*, **559**, pp. 68–73, 2013.

[8] Kang, J., Zhang, F.C., Yang, X.W., Lv, B. & Wu, K.M., Effect of tempering on the microstructure and mechanical properties of a medium carbon bainitic steel. *Materials Science and Engineering A*, **686**, pp. 150–159, 2017.

[9]   Zhao, P., Zhang, B., Cheng, C., Misra, R.D.K., Gao, G. & Bai, B., The significance of ultrafine film-like retained austenite in governing very high cycle fatigue behavior in an ultrahigh-strength MN-SI-Cr-C steel. *Materials Science and Engineering: A*, **645**, pp. 116–121, 2015.

[10]   Zhang, Z., Koyama, M., Wang, M.M., Tsuzaki, K., Tasan, C.C. & Noguchi, H., Effects of lamella size and connectivity on fatigue crack resistance of TRIP-maraging steel. *International Journal of Fatigue*, **100**, pp. 176–186, 2017.

[11]   Koyama, M., Zhang, Z., Wang, M., Ponge, D., Raabe, D. & Tsuzaki, K., Bone-like crack resistance in hierarchical metastable nanolaminate steels. *Science*, **355**, pp. 1055–1057, 2017.

[12]   Chai, G. & Zhou, N., Study of crack initiation or damage in very high cycle fatigue using ultrasonic fatigue test and microstructure analysis. *Ultrasonics*, **53**, pp. 1406–1411, 2013.

[13]   Zhao, P. et al., Non-inclusion induced crack initiation in multiphase high-strength steel during very high cycle fatigue. *Materials Science & Engineering A*, **712**, pp. 406–413, 2018.

[14]   Yajima, Z., Kishi, Y., Shimizu, K., Mochizuki, H. & Yoshida, T., Crack nucleation mechanism of austempered ductile iron during tensile deformation. *Materials Transactions*, **47**(1), pp. 82–89, 2006.

[15]   Shiozawa, K., Morii, Y., Nishino, S. & Lu, L., Subsurface crack initiation and propagation mechanism in high-strength steel in a very high cycle fatigue regime. *International Journal of Fatigue*, **28**, pp. 1521–1532, 2006.

[16]   Khan, M.K. & Wang, Q.Y., Investigation of crack initiation and propagation behavior of AISI 310 stainless steel up to very high cycle fatigue. *International Journal of Fatigue*, **54**, pp. 38–46, 2013.

[17]   Chai, G., The formation of subsurface non-defect fatigue crack origins. *International Journal of Fatigue*, **28**, pp. 1533–1539, 2006.

[18]   Peet, M. & Bhadeshia, H.K.D.H., www.msm.cam.ac.uk/map/steel/tar/mucg83.exe.

[19]   ASTM Standard E8/E8M-16a, *Standard Test Methods for Tension Testing of Metallic Materials*. ASTM International: West Conshohocken, PA, 2016.

[20]   Koistinen, D.P. & Marburger, R.E., A general equation prescribing the extent of the austenite-martensite transformation in pure iron-carbon alloys and plain carbon steels. *Acta Metallurgica*, **7**, pp. 59–60, 1959.

[21]   Cui, X.X., Northwood, D.O. & Liu, C., Role of prior martensite in a 2.0 GPa multiple phase steel. *Steel Research International*, **89**(10), p. 1800207, 2018.

[22]   Santofimia, M.J. & Zhao, L., Overview of mechanisms involved during the quenching and partitioning process in steels. *Metallurgical and Materials Transactions A*, **42**, pp. 3620–3626, 2011.

[23]   Toji, Y., Matsuda, H. & Raabe, D., Effect of Si on the acceleration of bainite transformation by pre-existing martensite. *Acta Materialia*, **116**, pp. 250–262, 2016.

CRITICAL PLANES CRITERIA APPLIED TO GEAR TEETH: WHICH ONE IS THE MOST APPROPRIATE TO CHARACTERIZE CRACK PROPAGATION?

FRANCO CONCLI & LORENZO MACCIONI
Faculty of Science and Technology, Free University of Bolzano, Italy

ABSTRACT

Tooth root bending fatigue is the most dangerous failure mode in gears. Indeed, it starts from the nucleation of a crack within the tooth root fillet region and the subsequent propagation up to the complete breakage of the tooth. To investigate this phenomenon, Single Tooth Bending Fatigue (STBF) tests are largely diffused. In these tests, an alternating bending stress at the tooth root is induced by the application of a pulsing force to the gear flank. In addition, this loading condition can be modelled through Finite Elements (FE) to study the stress state in the affected area. However, the application of strength criteria such as von Mises' can provide the equivalent stresses when the applied force reaches its maximum value but does not provide any insight in terms of fatigue behaviour. Nevertheless, the crack propagation can be investigated by analysing the results of FE analyses (which model the entire load cycle) through fatigue criteria based, for instance, on the critical plane concept. Previous studies conducted by the authors have shown that the different fatigue criteria (to study the tooth bending failure) lead to very different results. Therefore, the objective of the present paper is to compare the results of analyses carried out with different fatigue criteria based on critical plane, i.e. Findley, Matake, McDiarmid, Papadopoulos, and Susmel, with experimental outcomes, i.e. STBF tests on an aeronautical gears, to determine the most appropriate fatigue criterion to characterize the fatigue behaviour of these mechanical components. Results reveal that all fatigue criteria lead to consistent results when the target is to identify the most critical point. However, the Findley and Papadopulus criteria are found to be the most accurate for what concerns the evaluation of the damage. Among the others, Susmel turns out to be the most conservative criterion while the Findley criterion is the only one capable of identifying with good accuracy the direction of crack propagation.

Keywords: STBF, FEM, gears, fatigue, critical plane, crack.

1 INTRODUCTION

Nowadays, safety and reliability are undoubtedly Must-Be in the list of requirements for a gearbox and, therefore, the avoidance of failure due to the tooth root bending fatigue is a primary goal in gear design [1], [2]. With this respect, standards, e.g. ISO 6336-3 [3] and ANSI/AGMA [4], support gear design based on the determination of tooth bending strength. In particular, according to [3], the permissible bending stress is proportional to material strength that, in turn, is usually determined though experimental campaigns.

In the gear industry, the above-mentioned experimental campaigns are usually carried out by means of Single Tooth Bending Fatigue (STBF) tests [5]–[8] basically for economical and time reasons. In STBF tests, a gear made of the material and industrial processes to be tested is exploited as a specimen. Exploiting the Wildhaber property, two pulsating, competing, parallel and discordant forces are applied normal to two tooth flanks and tangent to the base circumference. In this way, using two simple anvils mounted on a universal testing machine, it is possible to load the teeth and to study the fatigue behavior experimentally [9], [10].

In order to study the stress state at the tooth root, experimental tests can be simulated numerically through Finite Element Models (FEM) [11], [12]. In this way, the force

WIT Transactions on Engineering Sciences, Vol 133, © 2021 WIT Press
www.witpress.com, ISSN 1743-3533 (on-line)
doi:10.2495/MC210021

imposed by the anvils can be translated in terms of stresses in the studied area [13]–[15]. To improve this calibration process, strain gauges measurements can be exploited [16]. However, although the FEM results provide relevant information on principal stresses, to obtain additional information on fatigue behavior these data require further elaboration [17].

Recent studies conducted by the authors, revealed that the results of FEM simulations can be analyzed through fatigue criteria based on the critical plane. Through this approach, it is possible to assert the criticality of the various positions within the fillet at the tooth root and the potential direction of propagation of a crack. However, results show a strong dependency on the fatigue criterion used.

The aim of this paper is to compare experimental results with numerical ones to shed a light on which could be the most appropriate fatigue criterion to study tooth root bending fatigue. In particular, a STBF test carried out in [15], [16] has been simulated by means of FEM and the results have been analyzed with different fatigue criteria based on the critical plane concept, i.e. Findley [18], Matake [19], McDiarmid [20], Papadopoulos [21], and Susmel et al. [22]. The outcomes of the elaboration have been compared with images of the fractures that occurred in the experimental campaign described in [15], [16].

2 BACKGROUND

A general time dependent stress tensor, referred to a specific point, can be represented by eqn (1). The maximum octahedral stress $\sigma_{h,max}$ (in the time window T) can be calculated according to eqn (2), where $\boldsymbol{\sigma_0}$ is a vector containing the principal stresses that, for the same time instant t, satisfies eqn (3), where $\bar{\bar{I}}$ is the identity matrix.

The stress vector $\boldsymbol{P_n}$ acting on a plane defined by a normal vector $\boldsymbol{n}(\phi_n, \theta_n)$ can be calculated through eqn (4). The modulus and the direction of $\boldsymbol{P_n}$ vary in time (Fig. 1(a)). In addition, $\boldsymbol{P_n}$ can be decomposed into a normal component $\boldsymbol{\sigma_n}$ (eqn (5)), having time-varying modulus and fixed direction, and a tangential component $\boldsymbol{\tau_n}$, having time-varying modulus and direction that, in turn, can be decomposed in its components aligned with the $\boldsymbol{u}$ and $\boldsymbol{v}$ directions respectively (eqn (6)) (Fig. 1(b)). $\boldsymbol{n}, \boldsymbol{u}, \boldsymbol{v}$ are defined as in eqn (7).

$$\bar{\bar{\sigma}}(t) = \begin{bmatrix} \sigma_{xx}(t) & \tau_{xy}(t) & \tau_{xz}(t) \\ \tau_{yx}(t) & \sigma_{yy}(t) & \tau_{yz}(t) \\ \tau_{zx}(t) & \tau_{zy}(t) & \sigma_{zz}(t) \end{bmatrix} \tag{1}$$

$$\sigma_{h,max} = \max_{T}\left\{\frac{1}{3}\sum_{i=1,2,3}\sigma_{0i}\right\} \tag{2}$$

$$\det\left|\bar{\bar{\sigma}}(t) - \boldsymbol{\sigma_0}\bar{\bar{I}}\right| = 0 \tag{3}$$

$$\boldsymbol{P_n}(\phi_n, \theta_n, t) = \bar{\bar{\sigma}}(t)\,\boldsymbol{n}(\phi_n, \theta_n) \tag{4}$$

$$\sigma_n(\phi_n, \theta_n, t) = \boldsymbol{n}^T(\phi_n, \theta_n)\bar{\bar{\sigma}}(t)\,\boldsymbol{n}(\phi_n, \theta_n) \tag{5}$$

$$\boldsymbol{\tau_n}(\phi_n, \theta_n, t) = \boldsymbol{u}^T(\phi_n, \theta_n)\bar{\bar{\sigma}}(t)\boldsymbol{u}(\phi_n, \theta_n) + \boldsymbol{v}^T(\phi_n, \theta_n)\bar{\bar{\sigma}}(t)\,\boldsymbol{v}(\phi_n, \theta_n) \tag{6}$$

$$\boldsymbol{n}(\phi_n, \theta_n) = \begin{bmatrix} \cos\phi_n \sin\theta_n \\ \sin\phi_n \sin\theta_n \\ \cos\theta_n \end{bmatrix}; \boldsymbol{u}(\phi_n, \theta_n) = \begin{bmatrix} -\sin\theta_n \\ \cos\phi_n \\ 0 \end{bmatrix}; \boldsymbol{v}(\phi_n, \theta_n) = \begin{bmatrix} -\cos\phi_n \cos\theta_n \\ -\sin\phi_n \cos\theta_n \\ \sin\theta_n \end{bmatrix} \tag{7}$$

For periodic stresses, P_n describes a closed curve in the space and, therefore, τ_n describes a closed curve in the plane. This curve is called as Γ_n in Fig. 1(b). With respect to σ_n, along the period T, it assumes different values from a minimum $\sigma_{n,min}$ to a maximum $\sigma_{n,max}$ (Fig. 1(b)). Therefore, it is possible to define the value of the alternating stress (acting on the plane having normal n) $\sigma_{n,a}$ according to eqn (8).

$$\sigma_{n,a} = \max_T\{\sigma_n(t)\} - \min_T\{\sigma_n(t)\} = \sigma_{n,max} - \sigma_{n,min} \qquad (8)$$

The curve Γ_n is representative of the tangential stresses acting on the studied plane during the entire loading cycle. To translate Γ_n into a value of alternate tangential stress $\tau_{n,a}$ (exerting on the plane with normal n) different methods can be found in the literature. The most diffused one is the Minimum Circumscribed Circle (MCC) (eqn (9)) [23], i.e. $\tau_{n,a}$ is considered as the radius of the smallest circle that can entirely contain the curve Γ_n (Fig. 2).

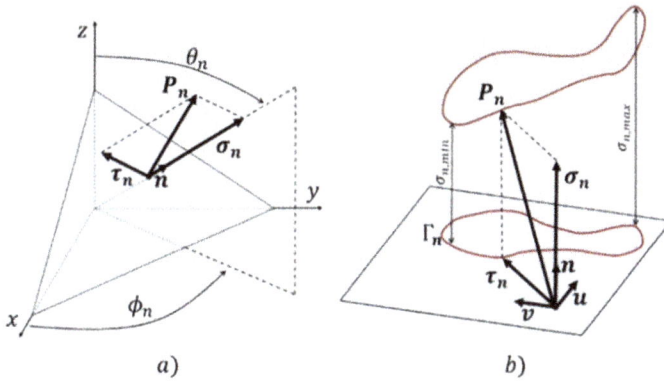

Figure 1:   (a) Components of $P_n(\phi_n, \theta_n, t)$ on the plane $n(\phi_n, \theta_n)$; and (b) definition of the curve Γ_n.

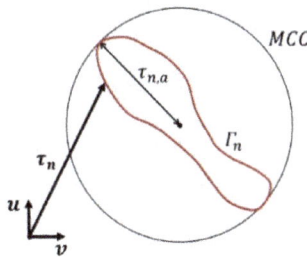

Figure 2: Minimum circumscribed circle (MCC) method.

$$\tau_{n,a} = MCC_T\{\tau_n(t)\} \qquad (9)$$

For each plane (having normal n), that can be defined by varying the parameters (ϕ_n, θ_n), it is possible to compute the related stress parameters, i.e. $\tau_{n,a}$, $\sigma_{n,max}$, and $\sigma_{n,a}$. In the present paper, for the critical plane, the corresponding spherical coordinates and the related stresses will be labelled with the subscript c, i.e. $\phi_c, \theta_c, \tau_{c,a}, \sigma_{c,max}$, and $\sigma_{c,a}$.

The damage parameter of each fatigue criterion based on the critical plane can be represented as in eqn (10), where the parameters k and f are constants related to the material properties and S is a variable related to the normal stresses exerting on the critical plane. A summary on how the k, f and S parameters are defined based on the different fatigue criteria can be found in eqns (11)–(15). In particular, it is possible to notice that for the calculation of the parameter k according to the McDiarmid criterion, the ultimate tensile stress σ_R is required. For the other fatigue criteria, k can be calculated through the material fatigue limit at symmetrical alternating bending loading (σ_f), and the material fatigue limit at symmetrical alternating torsional loading (τ_f). With respect to the variable S, the Papadopoulos criterion considers the maximum octahedral stress $\sigma_{h,max}$ while, the others, consider stresses related to the critical plane, i.e. $\sigma_{c,max}$ in Finley and McDiarmid, $\sigma_{c,a}$ in Matake, and $\sigma_{c,max}/\tau_{c,a}$ in Susmel et al. [22]

$$DP = \tau_{c,a} + kS \leq f \tag{10}$$

$$DP_{Findley} = \tau_{c,a} + \frac{2r_{\tau/\sigma}-1}{2\left(\sqrt{r_{\tau/\sigma}-r_{\tau/\sigma}^2}\right)}\sigma_{c,max} \leq \frac{\tau_f}{2\left(\sqrt{r_{\tau/\sigma}-r_{\tau/\sigma}^2}\right)} \tag{11}$$

$$DP_{Matake} = \tau_{c,a} + \left(2r_{\tau/\sigma}-1\right)\sigma_{c,a} \leq \tau_f \tag{12}$$

$$DP_{Susmel\ et\ al.} = \tau_{c,a} + \left(\tau_f - \frac{\sigma_f}{2}\right)\frac{\sigma_{c,max}}{\tau_{c,a}} \leq \tau_f \tag{13}$$

$$DP_{Papadopoulos} = \tau_{c,a} + \left(\frac{3}{2}\left(2r_{\tau/\sigma}-1\right)\right)\sigma_{h,max} \leq \tau_f \tag{14}$$

$$DP_{McDiarmid} = \tau_{c,a} + \frac{\tau_f}{2\sigma_R}\sigma_{c,max} \leq \tau_f \tag{15}$$

where

$$r_{\tau/\sigma} = \tau_f/\sigma_f \tag{16}$$

The determination of the critical plane (ϕ_c, θ_c) differs for the different fatigue criteria. According to the Findley criterion, the critical plane is the plane on which the damage parameter assumes its maximum value (eqn (17)) while, for the other fatigue criteria, the critical plane coincides with the one on which the $\tau_{c,a}$ reaches its maximum value (eqn (18)). Therefore, the application of the Findley criterion could lead to the identification of a critical plane having a different orientation with respect to the critical plane found applying the other fatigue criteria considered in this work. Eventually, the safety factor S_F can be defined as in eqn (19). $S_F > 1$ means that the analyzed point has not reached the critical value according to the studied criterion (and vice versa for $S_F < 1$).

$$(\phi_c, \theta_c) \rightarrow \max_{\phi,\theta}\{\tau_{n,a}(\phi,\theta) + kS(\phi,\theta)\} \tag{17}$$

$$(\phi_c, \theta_c) \rightarrow \max_{\phi,\theta}\{\tau_{n,a}(\phi,\theta)\} \tag{18}$$

$$S_F = \frac{f}{DP(\phi_C, \theta_C)} \tag{19}$$

3 MATERIAL AND METHOD

In the present paper, a gear geometry exploited in [22], [16] has been modelled in the STBF test condition. The gear parameters listed in Table 1 have been used as input in KISSsoft® to create a CAD model of the gear. Therefore, this model was imported into the open source FE software Salome-Meca/Code_Aster where the STBF test has been simulated.

Table 1: Geometrical parameter of the simulated gear.

Description	Symbol	Unit	Value
Normal module	m_n	(mm)	3.77301
Normal pressure angle	α_n	(°)	22.5
Number of teeth	z	(–)	32
Face width	b	(mm)	15
Profile shift coefficient	x	(–)	0.0681
Dedendum coefficient	h_{fP}^*	(–)	1.3153
Root radius factor	ρ_{fP}^*	(–)	0.36
Addendum coefficient	h_{aP}^*	(–)	1.1595

In the FEM simulation, symmetries were exploited to shorten the calculation time. In particular, a quarter of the gear has been meshed with an extruded grid. The quality of the mesh has been improved exclusively in the tooth subjected to the load. More specifically, the tooth in question have been meshed with hexahedral elements and the mesh density was increased in that region (Fig. 3). A non-linear simulation has been performed exploiting 40 time-steps (along the period T) for loading cycle. In Fig. 3, it is possible to see the faces in contact. The nodes of the anvil involved in the contact were 128 (105 hexahedral elements) while the nodes of the tooth flank involved in the contact were 248 (210 hexahedral elements). The radial symmetry (i.e. a plane containing the axis of the gear and parallel to the anvil contact face) was exploited to constrain the gear and a pulsating force varying sinusoidally from a minimum value of 3.7 kN to a maximum value of 37 kN was applied to the anvil to simulate the real testing condition. Indeed, this loading condition lead to the permissible bending stress for the studied gear according to [15], [16].

Typical steels properties have been applied to the components. The $\bar{\bar{\sigma}}(t)$ in the nodes within the tooth root fillet region have been extracted (Fig. 3). In the present study, 31 nodes discretize the studied area, i.e. each point where fracture could nucleate. Moreover, the anvil has been modelled with 2,048 nodes (1,575 hexahedral elements), the contacting tooth has been modelled through 30,256 nodes (25,200 hexahedral elements) and the rest of the gear has been modelled through 15,230 nodes (19,296 tetrahedral elements). The above-mentioned mesh has been fine-tuned through a sensitivity analysis.

At this point, according with [24], the approaches presented in the background section have been applied to the results of the FEM simulation. In particular, the material studied presents $\sigma_f = 1,400\ MPa, \tau_f = 1,100\ MPa$, and $\sigma_R = 2,700\ MPa$. Consequently, the parameters k and f of eqn (10) can be calculated for each fatigue criteria according to eqns (11)–(15).

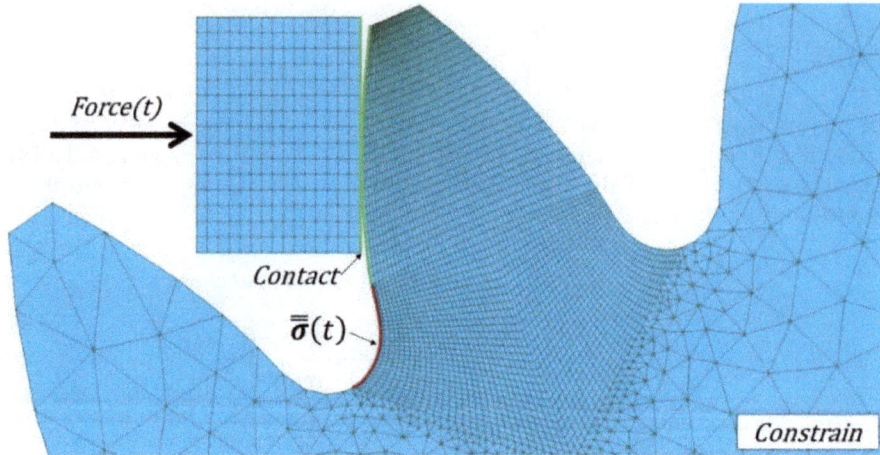

Figure 3: Finite element models of a STBF test.

Therefore, for each node within the tooth root fillet it is possible to study the stress tensor $\bar{\bar{\sigma}}(t)$ through the fatigue criteria presented in the background section. In particular, it is possible to individuate the critical plane for each node N $(\theta_c \phi_c|_N)$. This can be performed through eqns (17) and (18). The identification of the critical plane for each node allows achieving a twofold objective. First, it allows to evaluate the damage parameter (through eqn (10)) and, therefore, to calculate S_F (eqn (19)). In this way, it is possible to recognize the most critical node and to evaluate the differences between the various nodes in terms of probability of failure. Second, it allows identifying the direction of the initial propagation of the crack if it nucleates in any of the studied nodes. The determination of this direction was used, in comparison with the experimental ones to assess the effectiveness of each criteria to correctly predict the failure. Indeed, in STBF specimens, it is possible to identify both the point where the crack nucleated and the direction of crack propagation for each tooth that failed for tooth bending fatigue during the test.

In the present study, $\bar{\bar{\sigma}}(t)$ have been elaborated according to five different fatigue criteria i.e. Findley (eqn (11)), Matake (eqn (12)), Susmel et al. (eqn (13)), Papadopoulos (eqn (14)), and McDiarmid (eqn (15)). A Matlab routine have been implemented to explore all possible planes passing through each node with an angular resolution of $0.5°$ and identifying the critical planes. The results were compared with pictures of the STBF specimens to evaluate the capability of the various criteria to:

- identify the actual critical node;
- determine the actual crack direction; and
- provide a S_F consistent with the experimental measurements.

4 RESULTS AND DISCUSSION

A schematization of two cracks emerged during the experimental tests can be seen in Fig. 4. In particular, crack A is the recurrent one while crack B is nucleated in a different point probably due to the presence of defects in the material. This figure was created by observing the image of experimental specimens where the crack had broken the tooth or had propagated in its initial part. The actual images were not reported for confidentiality of the study. This schematization allows to:

- Identify the critical node, i.e. the intersection of crack A with the fillet; and
- Identify the direction of crack propagation at the critical node and at least one other node, e.g. crack B.

In Fig. 4, the angles between the direction of crack propagation and the tooth axis are shown. A coordinate χ that moves along the fillet has been defined. In Fig. 4, it is possible to see where this coordinate $\chi = 0$ and where $\chi = 1$ (values vary linearly along the profile of the fillet). Through this coordinate, it is possible to locate the critical node (e.g. $\chi = 0.40$ for crack A) and/or any point on the fillet at the tooth root (e.g. $\chi = 0.78$ for crack B). Such dimensions are obviously subject to measurement errors but are sufficiently accurate for the purposes of this paper.

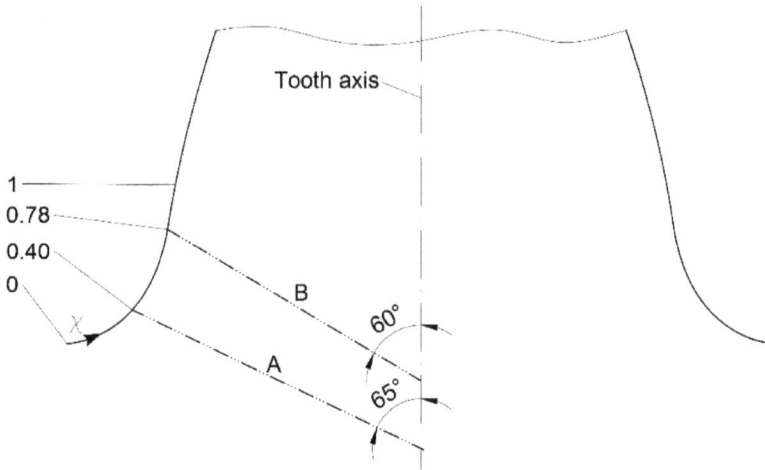

Figure 4: Schematization of experimentally detected cracks.

In Table 2, the S_F calculated according to different fatigue criteria are reported. In particular, the S_F has been calculated in the critical node, i.e. where the damage parameter is maximum (among the various nodes) for each criterion. More specifically, for the Findley and the Papadopoulos criteria the critical node is located at a coordinate $\chi = 0.39$ while for the other criteria the node $\chi = 0.42$. Therefore, it is possible to assert that all criteria identify the critical point with a variation less than 5%.

Table 2: S_F emerged through different fatigue criteria.

S_F according to				
Findley [18]	Matake [19]	Susmel et al. [22]	Papadopoulos [21]	McDiarmid [20]
1.08	1.96	0.79	1.13	2.14

Since the force set in the simulation leads to the permissible bending stress, S_F is expected to be equal to 1. Indeed, a lower force should lead to infinite life of the tooth (with the related induced stresses resulting below the fatigue limit). In Table 2, it is possible to notice that the criteria of Findley and Papadopoulos lead to S_F that approximate unitary values. Therefore, these fatigue criteria are the most appropriate to provide a S_F consistent with the experimental measurements. The Susmel et al. [22] criterion leads to the lowest S_F value and, therefore, it can be considered the most conservative one. The Matake [19] and McDiarmid [20] criteria lead to excessively high values of S_F and, therefore, they do not result to be appropriate to study tooth bending fatigue.

The graphical representation of the results according to the criteria of Findley and Papadopoulos are reported in Figs 5 and 6 respectively. This two-dimensional representation is made possible by the fact that all critical planes are perpendicular to the view shown. In the figures, the dash double-dotted segments represent the direction of the critical planes calculated for each node for the Findley (Fig. 5) and Papadopoulos (Fig. 6) criteria, i.e. the two criteria that provide more appropriate values of S_F. The length of each segment is proportional to the value of the damage parameter calculated on the relevant critical planes, i.e. the longest segment represents the plane on which the damage parameter is the maximum. The angle reported is exclusively for the critical plane passing through the critical node and the tooth axis. To simplify the identification of the critical plane having the highest damage parameter this has been made bold.

By comparing Figs 5 and 6, it is possible to notice how the Findley criterion leads to critical planes which form a greater angle with respect to the axis of the tooth e.g. 67° for the most critical plane. On the other hand, the Papadopoulos criterion, as well as the other criteria that identify the critical plane through eqn (18), lead to critical planes that tend to be less inclined with respect to the axis of the tooth e.g. 6° for the most critical plane.

In both Papadopoulos and Findley results, it is possible to observe a rapid variation in the angle of critical planes between some adjacent nodes. Applying Papadopoulos, this effect occurs at nodes near the origin of χ with a difference of about 85° between the two

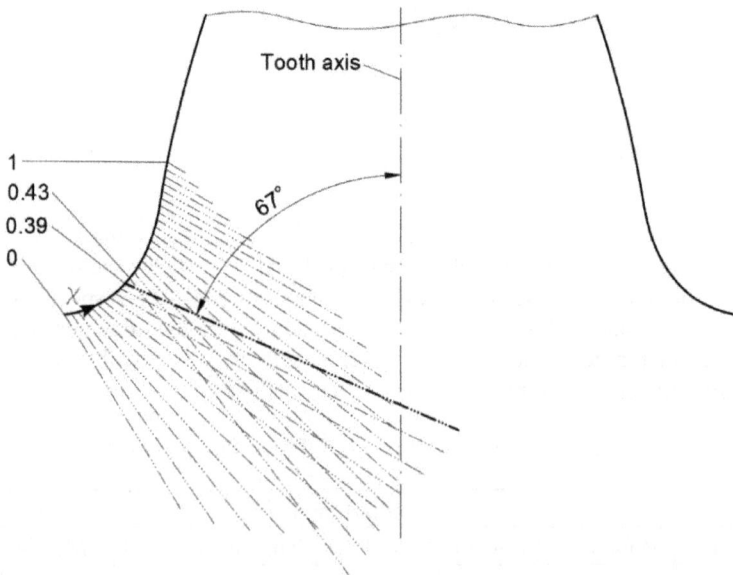

Figure 5:   Critical planes passing through the various nodes within the tooth root fillet according to the criteria of Findley.

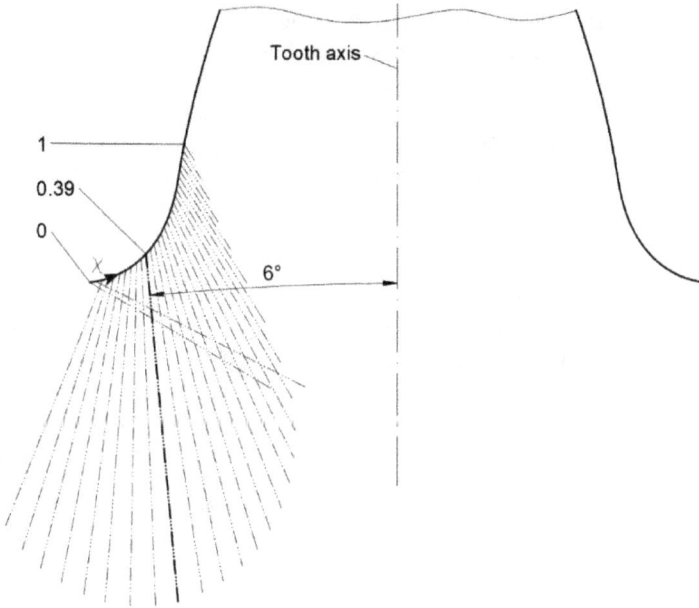

Figure 6:   Critical planes passing through the various nodes within the tooth root fillet according to the criteria of Papadopoulos.

critical planes. On the other hand, applying Findley, it occurs between the two nodes having coordinates $\chi = 0.39$ and $\chi = 0.43$ with a difference of about $30°$ between the two critical planes. This phenomenon can be attributed to the notching effect of the fillet radius.

Eventually, by comparing the numerical results (Figs 5 and 6) with the experimental ones (Fig. 4) it clearly emerges that for both cracks (A and B in Fig. 4) the Findley criterion succeeds in better representing the direction of crack propagation. Indeed, crack A (Fig. 4) shows an angle of $65°$ and a critical point of coordinate $\chi = 0.40$ and it is well represented by the crack emerged by applying Findley in $\chi = 0.39$ which lead to a crack having an angle of $67°$. The same can be observed for crack B (Fig. 4). Indeed, it shows an angle of $60°$ and a critical point of coordinate $\chi = 0.78$ and, in turn, it is well represented by the crack emerged by applying Findley in $\chi = 0.78$ that show an angle of around $58°$.

The same comparison conducted exploiting the results of Papadopoulos and/or the other criteria, leads to errors even much greater than $45°$.

5 CONCLUSIONS

In conclusion, criteria based on critical planes have been applied to the results of a FEM simulation (in terms of $\bar{\bar{\sigma}}(t)$) to evaluate the damage of various points of the tooth root fillet due to fatigue and to identify the direction of crack propagation in case the crack nucleated near to ones of the studied points. More specifically, the simulation was set to represent an experimental condition studied in [15], [16] and the applied force was calibrated to achieve a permissible stress level i.e. a stress level that lead to a unitary safety factor. Images of cracks obtained experimentally were compared with numerical results achieved by applying the different fatigue criteria. The different fatigue criteria were compared according to their ability to (1) evaluate a factor of safety close to unity, (2) identify the critical point where usually the crack nucleates, and (3) identify the direction of crack propagation. Results reveal that Findley and Papadopoulos criteria lead to a S_F very close to unity in the most critical node i.e. 1.08 and 1.13 respectively. In addition, it has emerged that the criterion of Susmel et al. [22], is the most conservative leading to $S_F = 0.79$ in the most stressed node. With respect to the individuation of the critical point, all the studied criteria reach the goal within a maximum variability of less than 5% of the arc length. Eventually, the fatigue criterion that better represents crack propagation is undoubtedly Findley's criterion.

Therefore, according to the results of this study, the Findley's criterion might be the most appropriate fatigue criterion for studying tooth bending fatigue. However, this statement is based on a single experimental validation conducted on a specific geometry and a specific material. Future studies conducted on different geometries and materials should be conducted to investigate the robustness of the insights obtained in this work.

REFERENCES

[1]   Fernandes, P.J.L., Tooth bending fatigue failures in gears. *Engineering Failure Analysis*, **3**(3), pp. 219–225, 1996. DOI: 10.1016/1350-6307(96)00008-8.

[2]   Hong, I.J., Kahraman, A. & Anderson, N., A rotating gear test methodology for evaluation of high-cycle tooth bending fatigue lives under fully reversed and fully released loading conditions. *International Journal of Fatigue*, **133**, p. 105432, 2020. DOI: 10.1016/j.ijfatigue.2019.105432.

[3]   ISO 6336-3:2006, *Calculation of Load Capacity of Spur and Helical Gears, Part 3: Calculation of Tooth Bending Strength*, Standard: Geneva, CH, 2006.

[4]   ANSI/AGMA 2001-D04, *Fundamental Rating Factors and Calculation Methods for Involute Spur and Helical Gear Teeth*, American Gear Manufacturers Association: Alexandria, 2004.

[5]   Benedetti, M., Fontanari, V., Höhn, B.R., Oster, P. & Tobie, T., Influence of shot peening on bending tooth fatigue limit of case hardened gears. *International Journal of Fatigue*, **24**(11), pp. 1127–1136, 2002. DOI: 10.1016/S0142-1123(02)00034-8.

[6]   McPherson, D.R. & Rao, S.B., Methodology for translating single-tooth bending fatigue data to be comparable to running gear data. *Gear Technology*, pp. 42–51, 2008.

[7]   Dobler, D.I.A., Hergesell, I.M. & Stahl, I.K., Increased tooth bending strength and pitting load capacity of fine-module gears. *Gear Technology*, **33**(7), pp. 48–53, 2016.

[8]   Concli, F., Tooth root bending strength of gears: Dimensional effect for small gears having a module below 5 mm. *Applied Science*, **11**, p. 2416, 2021. DOI: 10.3390/app11052416.

[9]   Gorla, C., Conrado, E., Rosa, F. & Concli, F., Contact and bending fatigue behaviour of austempered ductile iron gears. *Proceedings of the Institution of Mechanical Engineers, Part C: Journal of Mechanical Engineering Science*, **232**(6), pp. 998–1008, 2018. DOI: 10.1177/0954406217695846.

[10]   McPherson, D.R. & Rao, S.B., *Mechanical Testing of Gears*, ASM International: Materials Park, OH, pp. 861–872, 2000.

[11]   Concli, F., Austempered ductile iron (ADI) for gears: Contact and bending fatigue behavior. *Procedia Structural Integrity*, **8**, pp. 14–23, 2018. DOI: 10.1016/j.prostr.2017.12.003.

[12]   Bonaiti, L., Concli, F., Gorla, C. & Rosa, F., Bending fatigue behaviour of 17-4 PH gears produced via selective laser melting. *Procedia Structural Integrity*, **24**, pp. 764–774, 2019. DOI: 10.1016/j.prostr.2020.02.068.

[13] Gasparini, G., Mariani, U., Gorla, C., Filippini, M. & Rosa, F., Bending fatigue tests of helicopter case carburized gears: Influence of material, design and manufacturing parameters. *Fall Technical Meeting*, Vol. 369, ed. AGMA, American Gear Manufacturers Association (AGMA), pp. 131–142, 2008.

[14] Gorla, C., Rosa, F., Concli, F. & Albertini, H., Bending fatigue strength of innovative gear materials for wind turbines gearboxes: Effect of surface coatings. *ASME International Mechanical Engineering Congress and Exposition*, **45233**, pp. 3141–3147, 2012. DOI: 10.1115/IMECE2012-86513.

[15] Gorla, C., Rosa, F., Conrado, E. & Concli, F., Bending fatigue strength of case carburized and nitrided gear steels for aeronautical applications. *International Journal of Applied Engineering Research*, **12**(21), pp. 11306–11322, 2017.

[16] Gasparini, G., Mariani, U., Gorla, C., Filippini, M. & Rosa, F., Bending fatigue tests of helicopter case carburized gears: Influence of material, design and manufacturing parameters. *Gear Technology*, **68**, p. 76, Nov./Dec. 2009.

[17] Bonaiti, L., Bayoumi, A.B.M., Concli, F., Rosa, F. & Gorla, C., Gear root bending strength: A comparison between single tooth bending fatigue tests and meshing gears. *Journal of Mechanical Design*, pp. 1–17, 2021. DOI: 10.1115/1.4050560.

[18] Findley, W.N., A theory for the effect of mean stress on fatigue of metals under combined torsion and axial load or bending. *Journal of Engineering for Industry*, **81**(4), pp. 301–305, 1959. DOI: 10.1115/1.4008327.

[19] Matake, T., An explanation on fatigue limit under combined stress. *Bulletin of JSME*, **20**(141), pp. 257–263, 1977. DOI: 10.1299/jsme1958.20.257.

[20] McDiarmid, D.L., Fatigue under out-of-phase biaxial stresses of different frequencies. *Multiaxial Fatigue*, ASTM International, 1985.
DOI: 10.1520/STP36245S.

[21] Papadopoulos, I., A high-cycle fatigue criterion applied in biaxial and triaxial out-of-phase stress conditions. *Fatigue & Fracture of Engineering Materials & Structures*, **18**(1), pp. 79–91, 1995. DOI: 10.1111/j.1460–2695.1995.tb00143.x.

[22] Susmel, L., Tovo, R. & Lazzarin, P., The mean stress effect on the high-cycle fatigue strength from a multiaxial fatigue point of view. *International Journal of Fatigue*, **27**(8), 928–943, 2005. DOI: 10.1016/j.ijfatigue.2004.11.012.

[23] Papadopoulos, I., Critical plane approaches in high-cycle fatigue: on the definition of the amplitude and mean value of the shear stress acting on the critical plane. *Fatigue & Fracture of Engineering Materials & Structures*, **21**(3), pp. 269–285, 1998.
DOI: 10.1046/j.1460-2695.1998.00459.x.

[24] Concli, F., Fraccaroli, L. & Maccioni, L., Gear root bending strength: A new multiaxial approach to translate the results of single tooth bending fatigue tests to meshing gears. *Metals*, **11**(6), p. 863, 2021. DOI: 10.3390/met11060863.

STRUCTURAL MODELLING OF MULTILAYER SKIS WITH AN OPEN SOURCE FEM SOFTWARE

LORENZO FRACCAROLI[1], CARLO GORLA[2] & FRANCO CONCLI[1]
[1]Faculty of Science and Technology, Free University of Bolzano/Bozen, Italy
[2]Politecnico di Milano, Italy

ABSTRACT
The design process of a ski is characterized by a short time of development due to continuous advancements in the material science and in the manufacturing processes as well as in customer's requirements. Nowadays, the development process is very often still based on several physical prototypes and trials and Finite Elements Analysis (FEA) is a significant method to reduce times needed. The aim of this work is to develop a reliable numerical simulation of an existing mountaineering ski, able to predict the performance of the real element. For this purpose, an initial mechanical characterization of all the constituents used in the ski manufacturing was performed. Tensile tests in two directions were performed on flat bone-shaped samples laser cut from sheets. Combining the results of the tensile tests with Digital Image Correlation (DIC) data it was possible to approximate the four in-plane (XY) elastic properties, namely, the two elastic modules, the shear module and the Poisson ratio (E_x, E_y, G_{xy}, v_{xy}). The DIC free software used is GOM Correlate. Results of the combined "tensile tests – DIC" approach were after verified with FEM simulations reproducing the testing configuration. The digital model of the ski was created starting from the nominal geometry. The whole procedure of modelling, meshing and FE analysis was performed in the open source software Code_Aster/Salome-Meca. Using this kind of software, which code is free to use and modify, permits to reduce costs due to its free license. The real component was tested in a three-point bending and torsion test. This kind of experiments were replicated on the FEM model and results were compared. The comparison highlighted discrepancies of 2.5%–10% with respect to the real component.
Keywords: ski mountaineering, FEM, Code_Aster, DIC, composite materials, Salome-Meca.

1 INTRODUCTION

The current approach to the design of ski-mountaineering (skimo or skialp), and of those designed for racing purposes, is characterized by a relevant testing activity. Typically, several prototypes are manufactured, and a preliminary screening is based on the based on the experimental measurement of their basic mechanical properties, like torsional and bending behavior. After that, best prototypes are tested by professionals, who can give feedback on feelings and behaviors of the ski in real conditions. Even if, especially in the case of small companies, the prototyping process is very often quite efficient, it is evident that this trial and error procedure has significant disadvantages like the cost for manufacturing prototypes and the time needed for manufacturing and testing the sport equipment. The chance of reducing the necessary physical prototypes, based on a preliminary overview by means of virtual prototypes could represent an improvement of the whole design and manufacturing process. Simulations based on the use numerical techniques such Finite Element Analysis (FEA) represent a suitable solution for the scope under discussion. These numerical techniques have been used for the design phases of race-carving skis [1]–[4]. The mechanical properties and behavior of downhill component is highlighted in [5]–[7], however, the FEA approach is not yet used for development in the skimo sector. In this particular snow-sport it is very important to find a compromise between weight and in-operation behavior, this fact increases the difficulty during design phase of the component.

The objective of this research was to develop a numerical model of a mountaineering ski based on data corresponding to a model already on the market produced by a leader

WIT Transactions on Engineering Sciences, Vol 133, © 2021 WIT Press
www.witpress.com, ISSN 1743-3533 (on-line)
doi:10.2495/MC210031

manufacturing company. The real skis have been subjected to experimental bending tests as well as specimens of the different materials used in the ski manufacturing process. A preliminary material characterization phase took place. Classic dog bone shape specimens were produced thanks to laser cut technology. It has to be pointed out that the test procedures are not necessarily based on the standard methods defined for each class of materials because the aim of the tests was to determine the mechanical properties to be introduced in the models and not prepare a complete technical datasheet of the material. Specimens created from material sheets were tested (tensile test) in the two principal directions. In exception for some materials in which only one test in each direction was done (due to low constituents' availability), two tests were performed in each direction. In order to characterize materials the use of Digital Image Correlation (DIC) [8]–[15] and Campbell [16] hypothesis was mandatory. During the development of the numerical model different simplifications were introduced. To conclude, real tests results (torsion and bending) [15], [17] were compared with the one achieved with the FEM simulation.

2 MATERIALS AND METHODS

2.1 Ski overview

Skis are usually produced with a composite structure that foresees different layers. The core of the component is typically made by a very lightweight and flexible material like wood or honeycomb. The inner part is protected with multilayer constituents like carbon, basalt and glass fibers reinforced materials. These reinforcement components play the major role for the resistance and stiffness of the ski. Moreover, the ski includes materials not having structural purposes, like the lower and the top layer. These are used for other reasons: the upper part is for improving the quality of the decoration of the ski while the lower material is chosen in order to reduce the friction of the ski on the snow. It is evident that it is fundamental to find a balance between all constituents in order to reach the best configuration with respect of the performances of the ski. This objective is even more difficult to fulfil when the weight reduction is fundamental, as for the racing skies or for skies which are aimed at a high market level, in which a lightweight design must be obtained without affecting the downhill performances. In such conditions, solutions with an increased number of layers represent the most suitable solutions: it is evident that without the use of numerical simulations like FEA the development time increases hugely, moreover the knowledge of the manufacturer represent the unique available tool.

2.2 Material characterization

Since material properties were not known, the first phase of the research was dedicated to the characterization of mechanical properties of each constituent present in the real component. Skies considered are built up with different layers of composites materials with a central wood softcore. Composites materials have two major constituents, fibers and matrix. Fibers take a major part of the load during operation while matrix must keep fibers together and protect them from external agents. For describing the mechanical behaviour of composites materials, the theory of orthotropic laminas and laminated structures has been used. The material was considered having just one principal direction. Orthotropic materials can be described in 3D by means of nine independent elastic constants, compared with the 21 required for a completely non isotropic material. For thin laminas that has only continuous fibers, behaviour can be described with four independent elastic constants: the two elastic modules, the in-plane shear module and one of the two in-plane Poisson's ratios. In this research, the two elastic modules were obtained directly from unidirectional tensile tests and the Poisson's ratio was extrapolated from Digital Image Correlation (DIC) measurements. The shear module was estimated with the approximated Campbell equation [16].

Materials are recognized by means of a code number. Specifically, 9, 24, 139, 207, 290, 135, 31, 86 and 217. Monoaxial tests were performed on an STEPLab UD04 (Fig. 1) testing machine capable to apply static forces up to 4.5 kN. Tests crosshead speed was set at 0.1 mm/min. While the elastic modules can be directly obtained from the tensile tests, the estimation of the Poisson's ratio requires the calculation of the strains in two directions (eqn (1))

$$v_{xy} = -\frac{\varepsilon_y}{\varepsilon_x} \; ; \; v_{yx} = -\frac{\varepsilon_x}{\varepsilon_y}. \tag{1}$$

One option could be the adoption of multiaxial strain gauges. However, Digital Image Correlation (DIC) represents a valid alternative. One camera (reflex Nikon D750 with a 24–85 zoom and a stabilizer), acquires pictures during timestep. A correct illumination was guaranty by a dedicated lamp source.

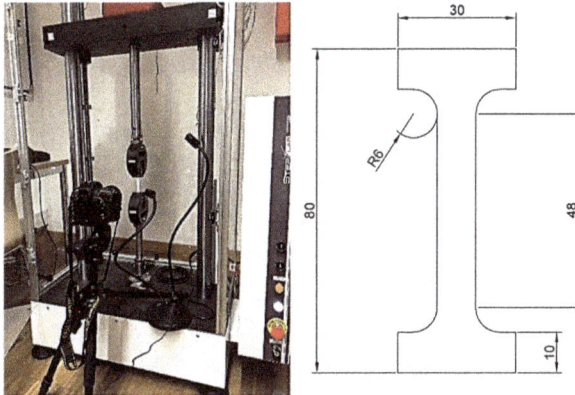

Figure 1:   STEPLab UD04 tensile machine and testing setup; (right) dog-bone sample geometry and dimensions.

2D Digital Image Correlation is an advanced optical measurement technique that allows to reconstruct displacement and strain fields of an inspected component having a planar surface. This technology is based on the greyscale; therefore, a characteristic "speckle" pattern based on the gray scale must be made on the inspected surface before testing. The camera should be positioned perpendicularly to the specimen surface in order to avoid out of plane measurements. The camera acquired image during the tensile test with a constant time step. The first picture acquired (undeformed sample) is used as reference. The post-processing algorithm creates a virtual grid on the image of the specimen's surface dividing the total area into several smallest areas called subsets or facets. Cross correlation operations allow the recognition of facets during each step (equivalent to diverse picture and displacements) (Fig. 2). When all facets are recognized at each time, the algorithm reconstructs the displacement field and successively the strain field of the tested component. The package used is GOMrrelate [18].

WIT Transactions on Engineering Sciences, Vol 133, © 2021 WIT Press
www.witpress.com, ISSN 1743-3533 (on-line)

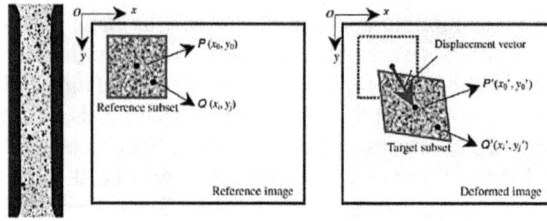

Figure 2: DIC; (left) surface pattern; (right) cross correlation.

In this manner it is possible to measure strains in the two different directions. With these new data is now possible to calculate the Poisson ratio with (eqn (1)), while the shear module is computed with the Campbell equation (eqn (2))

$$G_{xy} = \frac{E_x}{(1+v_{xy})} + \frac{E_y}{(1+v_{yx})}. \tag{2}$$

2.3 FEA

In this work the open source computer software Code_Aster/Salome-Meca was used. It is an open source Finite Element (FE) environment developed by EDF (Électricité de France). In Fig. 3 is visible the load configuration of the tested ski.

Two reproductions with different levels of simplifications were tested.

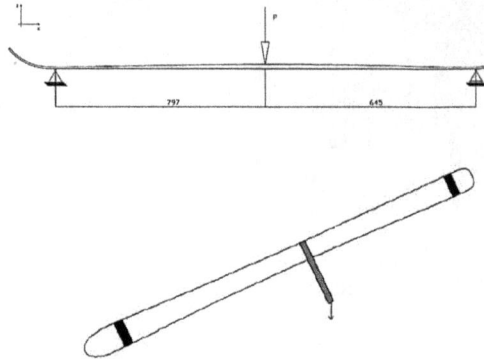

Figure 3: Load configuration: the three-point bending (up) and torsion-bending tests (bottom) – reference case for the FE simulations.

2.3.1 Shell-model

Considering the configuration of the three-point bending test, some geometrical considerations were made. First, it was decided to exploit the z–x plane symmetry of the ski, therefore only one half of the structure was modeled. Secondly, tail and top were cut off in order to consider only the component that lay between supports. In addition, also steel laminas were neglected due to their very low structural contribution. Fig. 4 shows the shell model. All the small subdivision that can be seen were necessary for correctly defining the height of the ski. The shell modelling in Code_Aster/Salome-Meca does not recognize curved surfaces, therefore at each small area was assigned the correspondent value of thickness. The component was separated into 145 parts in order to have a better estimate of the curvature of the component. The final grid consists in 22,382 quadrangular mesh elements. The torsion-bending test was simulated numerically with a similar model, without exploiting the z–x plane-symmetry.

Figure 4:   Shell models used for the three-point-bending (left) and torsion-bending (right) simulations.

2.3.2 Solid-model

The 3D modelling foresees to exploit the same simplifications used for the 2D analysis. The same geometrical simplifications of the shell-model were used also for the 3D simulations. Moreover, other simplifications were necessary for defining correctly material parameters. As explained in Section 2.1, for describing composites material in the three-dimensional case are necessary nine elastic constants. According to Aerospaziali [19], if materials are composed by only unidirectional long fibers, it is possible to assume that elastic modules in the radial direction are the same. Moreover, tests were done on low thickness laminas, this fact does not permit to follow standards for computing the outstanding out of plane properties. For this reason, Module G and Poisson's ratio were set equal to the one obtained previously with the in-plane (eqn (1)) and (eqn (2)) considerations. Assumptions taken were the following: Poisson's ratios and shear modules were set equal in each of the three principal directions while the third elastic module was set equal to the one in the y direction. On the other hand, as shown in Figs 5 and 6, in the solid model the lateral protective layer was also considered.

Figure 5: Isometric view of the solid model used for the three-point-bending test.

Figure 6: Different layers of the solid model.

Three finite elements in the thickness were created for each material layer to correctly describe the stress profile. The grid was formed by quadratic hexahedrons. A total mesh of 500 k cells was created for the three points bending model.

3 RESULTS

The results of the tensile test in terms of stress strain curves and numerical values are presented in Figs 7–13 and Table 1 respectively.

Longitudinal direction

Transversal direction

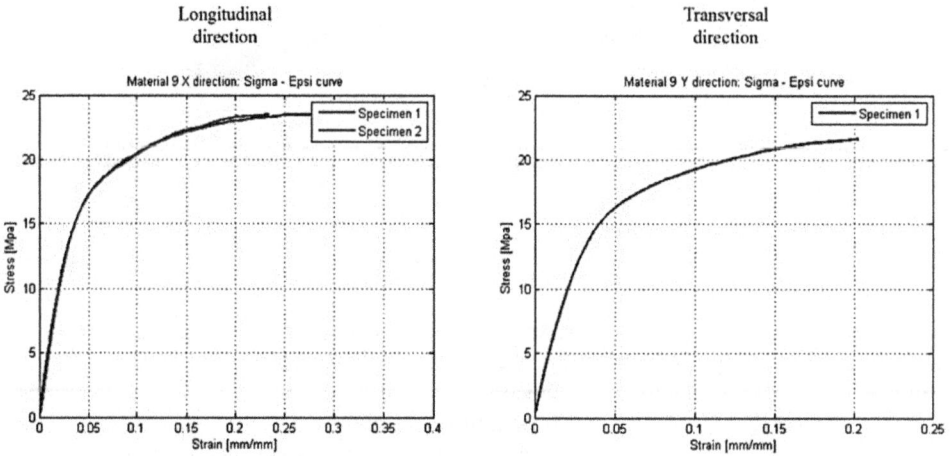

Figure 7: Material 9 tensile tests.

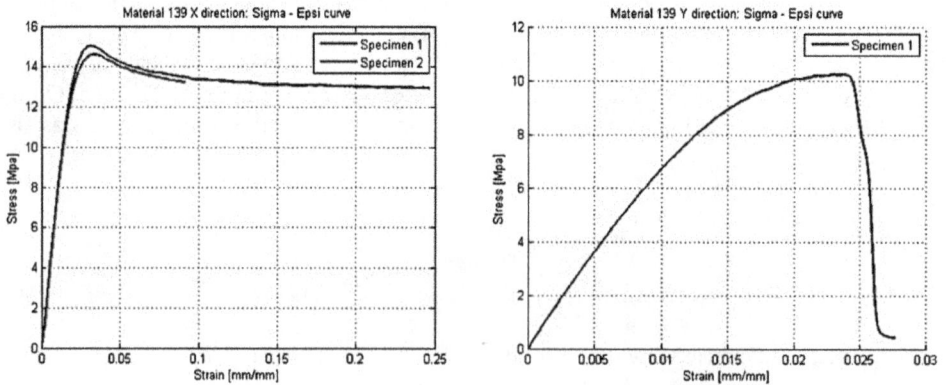

Figure 8: Material 139 tensile tests.

Table 2 shows the higher displacement of the center of the ski when subjected to a load of 120 N in the three-point bending tests. Tests were performed with the help of two supports and a kettlebell for applying the load. Ski was blocked to the supports thanks to two clamps and the kettlebell positioned on the ski through its handle. Results for both experimental and numerical approach were confronted. Table 3 reports the results of the torsion-bending tests and simulations.

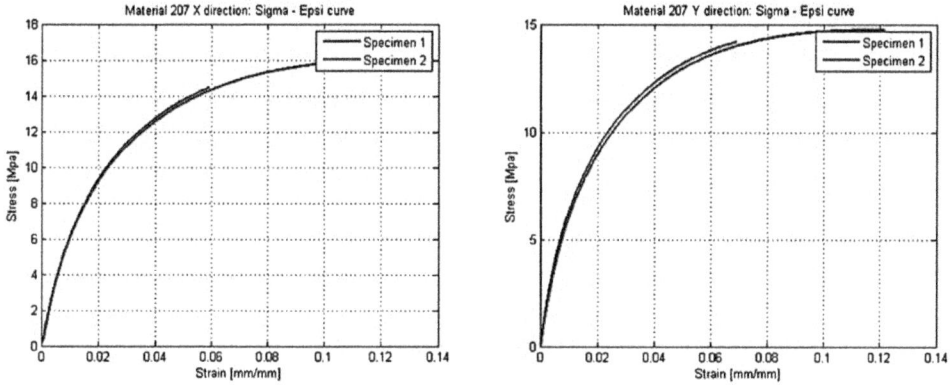

Figure 9: Material 207 tensile tests.

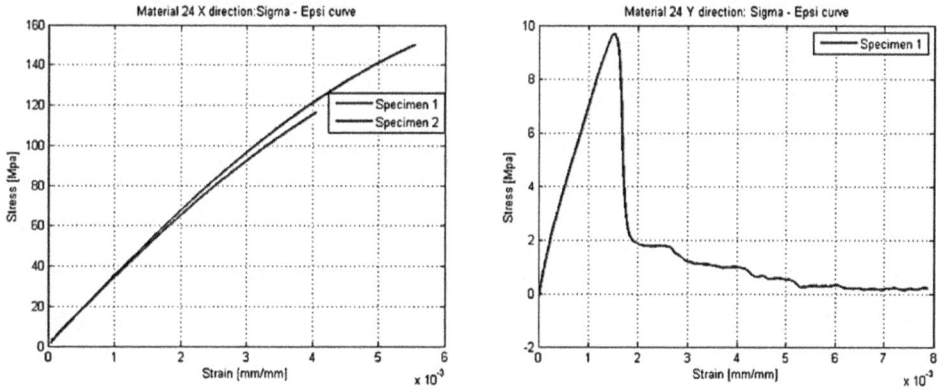

Figure 10: Material 24 tensile tests.

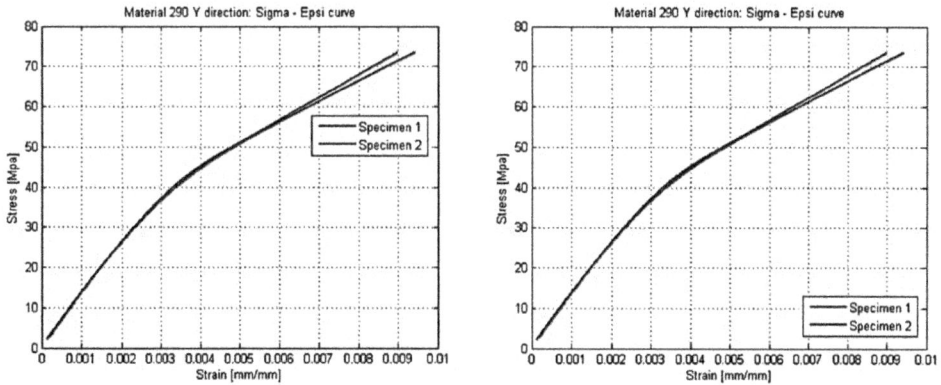

Figure 11: Material 290 tensile tests.

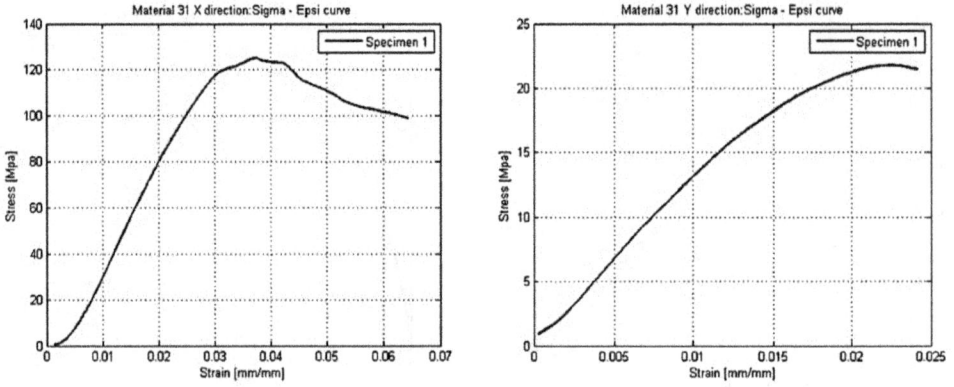

Figure 12: Material 31 tensile tests.

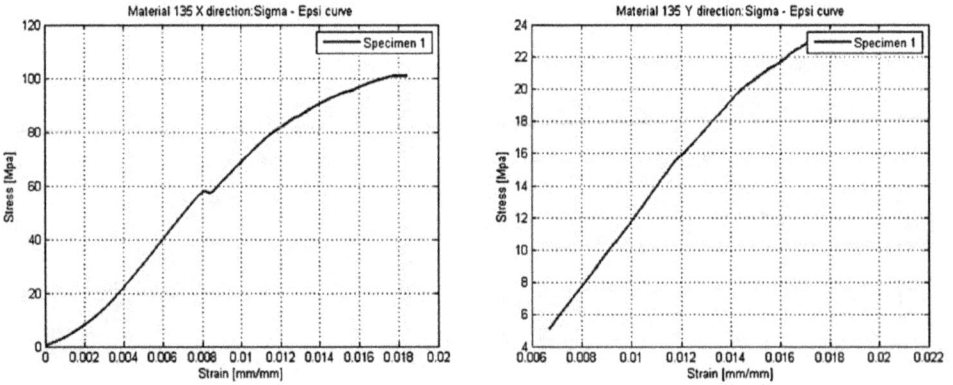

Figure 13: Material 135 tensile tests.

Table 1: In-plane properties of the tested materials.

	Material								
Material #	9	24	139	207	290	135	31	86	217
E_x (MPa)	550	35,000	800	850	20,000	4,000	2,000	13,700	35,000
E_y (MPa)	567	8,700	750	897	13,700	569	127	420	12,436
v_{xy} (-)	0.45	0.23	0.45	0.42	0.18	0.39	0.39	0.40	0.56
v_{yx} (-)	0.43	0.35	0.38	0.30	0.12	0.39	0.39	0.40	0.48
G_{xy} (MPa)	776	35,010	1,097	1,291	29,125	3,009	1,184	10,086	25,327

Table 2:   Comparison of the maximum deflection of the ski in the three-point bending test: FEM vs. experiments.

Model	Displacement (mm)	Simulation time* (min)	Error (%)
Experimental	40		
Shell	41	35	2.5%
Solid	44	60	10.0%
*on a 9.6 GFLOPS workstation			

Table 3:   Comparison of the maximum deflection of the ski in the torsion-bending test: FEM vs. experiments.

Model	Rotation angle (°)	Simulation time* (min)	Error (%)
Experimental	4.44		
Shell	4.61	35	3.0%
Solid	4.86	60	9.5%
*on a 9.6 GFLOPS workstation			

Fig. 14 reports the normal stresses in the point in which load was applied with respect to the thickness of the ski (0 match with the lower part) during the bending test. Fig. 15 is similar but refers to the bending-torsion test.

4 DISCUSSION

Considering the nature of the ski, only the linear part of the constitutive law is of interest. Most of the tested materials does not present discrepancies in elastic field. For material #9 #24 and #139, it can be appreciated that the two repetitions in the X direction do not show differences in the linear part. In the Y direction, because off the low presence of constituents, only one probe was tested. Considering materials #207 and #290 it is possible to observe that, in both directions, no differences are present into the elastic field between the different repetitions. For materials #135 and #31 a different approach was used. In these particular

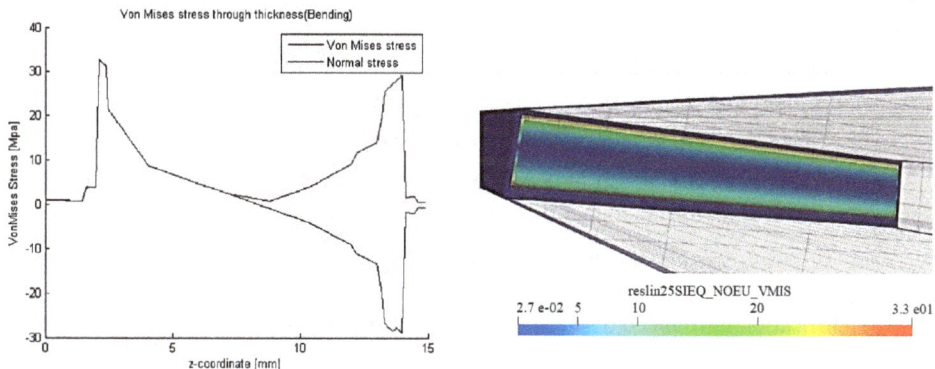

Figure 14:   Solid model results – three-point-bending test (normal and Von Mises stresses in the section).

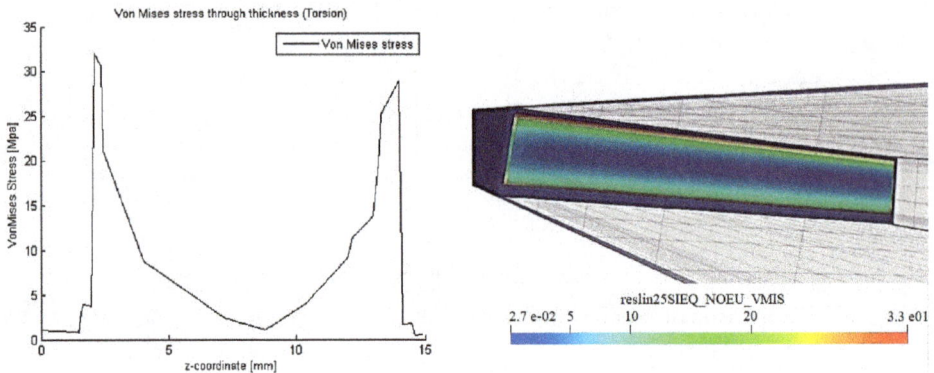

Figure 15: Solid model results – torsion-bending test (normal and Von Mises stresses in the section).

specimens it was impossible to create the classic dog bone probe, therefore a rectangular tester was obtained cutting the initial sheet of material. For these constituents, since the fibres were easily identifiable, the thickness and width of single filaments were measured. Consequently, the resistant area was obtained by multiplying the single fibres area by the total number of filaments present in the tested probe. The tests on the single material will be object of further refinement, but in this phase of the research, which is more focused on the definition of FEM models than on the exact characterization of the materials, the data collected seem enough.

FEM results shows a good agreement with the experimental measurements for both the modelling approaches (2D and 3D). The 2D analysis highlighted a maximum displacement (three-point-bending) of 41 mm with a 2.5% difference with respect to the experimental measure while the solid analysis gave as results equal to 44 mm (10% difference vs. experiments). In the torsion-bending analysis, the shell model predicts a rotation of 4.60° while the solid model a rotation of 4.86. The error with respect to the experimental measure is equal to 3% and 9.5% respectively. For the shell model, discrepancies in the predicted displacement/rotation with respect to the measured one can be attributed to the following reasons:

- Component was divided in small areas in which were assigned different values of heights, therefore the curvature of the ski is approximated and not correctly replicated.
- The steel edges and the lateral layers (cage) were neglected.

On the other side, comparing the solid model with the experimental test, it is possible to give some motivation for the higher discrepancy including:

- The ski was considered straight without considering its shape in the longitudinal direction.
- Simplification introduced during the material characterization phase may affect results

Differences between the solid model and the shell one is also due to the lateral cage. Both configurations have equal width, but the cage were simulated only in the 3D approach. Bending contribution of these lateral constituents component is lower with respect to the main core part of the ski. From Figs 14 and 15 it can be appreciated that some constituents do not give resistant contribution. This fact is highlighted considering materials at the border of the ski (bottom and top layer). Although their displacement, it is possible to see that they are not loaded. In fact, their contribution is for example to reduce the friction with the ground (bottom layer) or improve the insulation of the component form external agents that may degrade the internal layers. FEA allows an accurate evaluation on how each layer is contributing to the total stiffness of the ski.

These considerations are valid for both models (Three-point bending and torsion-bending tests). However, for the last mentioned it is possible to give a further consideration for explaining results discrepancies. Generally, for torsion the shear module G plays a pivotal role, it is possible that the fact that it was approximated and not directly computed may influenced results.

5 CONCLUSIONS

An accurate mathematical description of a real ski is still difficult to achieve. However, in this work a 2.5%–10% error approximation was reached. The difficulty to describe properly materials that were composing the sandwich structure influenced hugely the model replication. The fact that materials dimensions do not permit to follow standards has to be considered when evaluating results. The elastic module was calculated for each constituent (For material #86 and #217 properties are available in literature). Considering the results of the final simulations, it is possible to achieve satisfactory levels of approximation. The use of DIC was mandatory for a complete material characterization. The comparison between the real model and FEM simulations of three-point bending and torsion-bending tests was made. Better results were obtained with the shell approach. The solid model has a higher percentage error due to material parameters simplifications. However, theoretical predictions of the stress state and the behavior of the ski when subjected under bending were confirmed by experimental test. For the torsion-bending shell model, a maximum error of the 3% was found while for the solid one a 9.5% discrepancy was recorded. As for the bending case the plane model presents more reliable results. This fact is due to the lower level of approximation used in the 2D problem.

Once the model is validated it will be possible to easily change materials properties and layers dimensions having a virtual comparison between different design solutions. This fact can lead to a reduction of the time needed for ski development. Moreover, the need to produce several physical prototypes is not more necessary, as consequence it will be possible to decrease research and development costs. Future improvements will be a more accurate modelling of the ski without simplification in terms of material properties. The fact that steel edges were not modelled, influenced securely the performance of the ski.

It must be pointed out that in the real application the effective load conditions on the ski are much more complex and difficult to determine. The loads are applied in all the directions and the reaction coming from the contact with the snow are variable and strongly affected by the deflections of the ski and by the deformation of the snow. Moreover, the behaviour of the snow itself is widely variable and influenced by many parameters (temperature, humidity, time from the deposition, thermal and mechanical cycles experienced, etc.).

For this reason the validation of the simulation of ski under real load condition represents a hard job and also models of the snow should be considered, validated experimentally compared with simulations in a FE environment in order to verify the accuracy of the FEM ski model. These issues will be investigated in the next steps of the research. In conclusion it should be highlighted that while the results achieved are not perfectly reproducing the real ski performance, the accuracy is satisfactory considering all the simplifications and hypothesis introduced.

REFERENCES

[1]   Wolfsperger, F., Szabo, D. & Rhyner, H., "Development of alpine skis using FE Simulations," *Procedia Eng.*, **147**, pp. 366–371, 2016. DOI: 10.1016/j.proeng.2016.06.314.

[2]   Federolf, P., Roos, M., Lüthi, A. & Dual, J., Finite element simulation of the ski-snow interaction of an alpine ski in a carved turn. *Sport. Eng.*, **12**(3), pp. 123–133, 2010. DOI: 10.1007/s12283-010-0038-z.

[3]   Zboncak, R., Experimental verification of ski model for finite element analysis. *Conf. Experimental Stress Anal.*, **56**, pp. 450–456, 2018.

[4]   Mössner, M., Innerhofer, G., Schindelwig, K., Kaps, P., Schretter, H. & Nachbauer, W., Measurement of mechanical properties of snow for simulation of skiing. *J. Glaciol.*, **59**(218), pp. 1170–1178, 2013. DOI: 10.3189/2013JoG13J031.

[5]   Nordt, AAGSSLPK, Computing the mechanical properties of alpine skis. *Sport. Eng.*, **2**(2), p. 65, 1999. DOI: 10.1046/j.1460-2687.1999.00026.x.

[6]   Hirano, Y. & Tada, N., Mechanics of a turning snow ski. *Int. J. Mech. Sci.*, **36**(5), pp. 421–429, 1994. DOI: 10.1016/0020-7403(94)90045-0.

[7]   Cresseri, S. & Jommi, C., Snow as an elastic viscoplastic bonded continuum: A modelling approach. *Ital. Geotech.*, **4**, pp. 43–58, 2005.

[8]   Musotto, Z., Digital Image Correlation : Correlation applicazione di tecniche convenzionali e sviluppo di soluzioni la stima e l ' incremento dell ' accuratezza, 2012.

[9]   Crammond, G., Boyd, S.W. & Dulieu-Barton, J.M., Speckle pattern quality assessment for digital image correlation. *Opt. Lasers Eng.*, **51**(12), pp. 1368–1378, 2013. DOI: 10.1016/j.optlaseng.2013.03.014.

[10] Makeev, A., He, Y., Carpentier, P. & Shonkwiler, B., A method for measurement of multiple constitutive properties for composite materials. *Compos. Part A Appl. Sci. Manuf.*, **43**(12), pp. 2199–2210, 2012. DOI: 10.1016/j.compositesa.2012.07.021.

[11] Kowalczyk, P., Identification of mechanical parameters of composites in tensile tests using mixed numerical-experimental method. *Meas. J. Int. Meas. Confed.*, **135**, pp. 131–137, 2019. DOI: 10.1016/j.measurement.2018.11.027.

[12] Schreier, H.W. & Sutton, M.A., Systematic errors in digital image correlation due to undermatched subset shape functions, pp. 303–310.

[13] Wattrisse, B., Chrysochoos, A. & Muracciole, J., Analysis of strain localization during tensile tests by digital image correlation. pp. 29–39, 2000.

[14] Peters, W.H. and Ranson, W.F., Digital image techniques in experimental stress analysis. *Opt. Eng.*, **21**(3), pp. 427–431, 1982.

[15] Górszczyk, J., Malicki, K. & Zych, T., Application of digital image correlation (DIC) method for road material testing. *Materials (Basel)*, **12**(15), p. 2349, 2019. DOI: 10.3390/ma12152349.

[16] Yokoyama, T. & Nakai, K., Evaluation of in-plane orthotopic elastic constants of paper and paperboard. *Proc. SEM Annu. Conf. Expo. Exp. Appl. Mech*, **3**(2007), pp. 1505–1511, 2007.

[17] Fraccaroli, L. & Concli, F., Introduction of open-source engineering tools for the structural modeling of a multilayer mountaineering Ski under operation. *Appl. Sci.*, **10**(15), 2020.

[18] www.gom.com.

[19] Aerospaziali, T.E.M., CAPITOLO 32 - materiali compositi: la legge costitutiva ortotropa.

BENDING FATIGUE STRENGTH OF SMALL SIZE 2 MM MODULE GEARS

FRANCO CONCLI & LORENZO FRACCAROLI
Libera Università di Bolzano/Bozen, Piazza Università, Italy

ABSTRACT

Nowadays the use of gears has significantly increased due to new materials available and to the high variety of possible applications. The necessity to investigate the gears behavior in different conditions plays a pivotal role for the evolutionary process of this so widely used mechanical component. In addition, the miniaturization process that is still growing in most of the engineering fields brings the necessity to accurately analyze the material properties, and fatigue behavior of gears having reduced dimensions. In this framework, in this work an experimental analysis combined with theoretical calculation was used for describing the fatigue behavior of 2 mm module gears made of 39NiCrMo3. Single Tooth Bending Fatigue (STBF) tests were performed according on a tensile testing machine. Since the machine grippers were not suited for fatigue tests on gears, a dedicated test-tool was appositively manufactured to exploit the Wildhaber W5 property. Tests were performed at different loading levels following the stair-case approach. Test were performed with a ΔF of 100 N. The fatigue limit was computed using the statistic Dixon approach that allows a precise calculation of the fatigue limit also with a reduced number of experimental data. The results in terms of measured forces were converted into the corresponding stresses using the ISO 6336 approach. Thereafter, the results were compared with the one stated in the ISO standard for the same steel material.

Keywords: gears, STBF, 39NiCrMo3, fatigue.

1 INTRODUCTION

Nowadays, the use of miniaturized components is still growing in the mechanical sector. Hence, the use of 2 mm or less modules for the design of small gearboxes has significantly increased due to the various possible applications, like robotics, high power density transmission systems and all the applications related to the automotive sector. In this important field of mechanic, it was also demonstrated that the use of small sized gears can reduce pollution [1]. All these facts lead to the necessity to correct investigate the material behavior when used for manufacturing small gears. In particular, this work is aimed in highlighting the steps necessary to compute the material properties, namely the admissible root tooth bending stress σ_{Flim}, and to report the results for a 39NiCrMo3 steel. In literature several standards are available for the design of gears, such as the EU ISO 6336 [2], the German DIN 3990 [3] and the US ANSI/AGMA 2001-D04 [4]. However, such standards are based on data obtained on 5 mm module gears. Several works on gears having a normal module $m_n \geq 5$ [5]–[9] have highlighted the inaccuracy of the actual standards in predicting the fatigue strength. In particular it was shown that the size of the gear significantly affects the resistance of the solution. A research made by Steutzger [10], showed that for what concern the tooth root bending load-carrying capacity, the adoption of gears with normal module $m_n > 5$ is unfavorable. This evidence was confirmed by other scholars [11]–[13]. The tooth root bending behavior of big sized gears (nitrided and carburized) was already investigated by different scholars [14]–[16].

The gears behavior is influenced by several factors like the heat treatment, the surface finishing, temperature and the effective geometry of the gear, just to highlight some of the most important ones. Recently, some works in which small sized gears were investigated have been published [17], however this it is not still enough for having reliable data for the design phase of gears that have a normal module $m_n < 5$. The ISO 6336 method B provides an approach which relies on the comparison between the permissible stress σ_{FP} and the effective stress acting on the tooth σ_F. According to the standard, the two stress components are calculated according to eqns (1) and (2).

$$\sigma_{F0} = \frac{F_t}{b \cdot m} \cdot Y_F \cdot Y_S \cdot Y_\beta \cdot Y_B \cdot Y_{DT}, \tag{1}$$

$$\sigma_F = \sigma_{F0} \cdot K_A \cdot K_V \cdot K_{F\beta} \cdot K_{F\alpha}. \tag{2}$$

Considering the geometry specification and the application of the gear inspected Y_β, Y_{DT} and Y_B were set equal to 1 reducing the formulation at eqn (3)

$$\sigma_F = \frac{F_t}{b \cdot m} \cdot Y_F \cdot Y_S \cdot K_A \cdot K_V \cdot K_{F\beta} \cdot K_{F\alpha}, \tag{3}$$

where the form factor Y_F is defined as eqn (4) and the stress correction factor Y_S as eqn (5).

$$Y_F = \frac{\frac{6 h_{Fe}}{m} \cdot \cos\alpha_{Fen}}{\left(\frac{s_{Fn}}{m_n}\right)^2 \cdot \cos\alpha_n}. \tag{4}$$

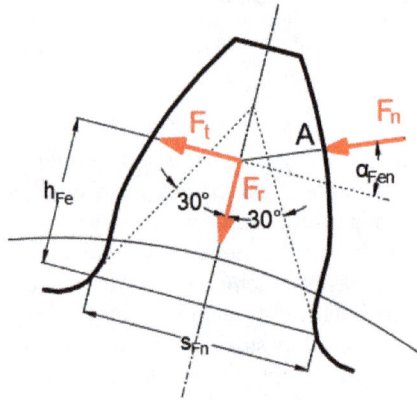

Figure 1: Representation of geometrical tooth parameters necessary for the computation of the Y_F factor.

$$Y_S = (1.2 + 0.13 \cdot L) \cdot q_s^{\frac{1}{1.21 + \frac{2.3}{L}}}. \tag{5}$$

With L defined as eqn (6) and the notch sensitivity q_s computed thanks to eqn (7)

$$L = \frac{s_{Fn}}{h_{fe}}, \tag{6}$$

$$q_s = \frac{s_{Fn}}{2 \cdot \rho_F}. \tag{7}$$

The application factor K_A depends on the applications because it considers the external loading effects. K_V, also called, dynamic factor takes into account internal dynamics loads while the last two factors $K_{F\alpha}$ and $K_{F\beta}$ are used to manage transvers and uneven loads acting on tooth during contacts. These phenomena can be caused by the deflection of the component during operation ore by manufacturing errors.

As stated, form ISO 6336 method B, the allowable stress of the material defined as σ_{FP} can be determined and compared with the one acting on tooth σ_F. σ_{FP} is described by eqn (8)

$$\sigma_{FP} = \sigma_{Flim} \cdot Y_{ST} \cdot Y_{NT} \cdot Y_{\delta relT} \cdot Y_{RrelT} \cdot Y_X. \tag{8}$$

By comparing eqns (8) and (3) it is possible to compute σ_{Flim} with eqn (9)

$$\sigma_{Flim} = \frac{Ft}{b \cdot m} \cdot \frac{Y_F \cdot Y_S}{Y_{ST} \cdot Y_{NT} \cdot Y_{\delta relT} \cdot Y_{RrelT} \cdot Y_X} \tag{9}$$

Y_{ST} and Y_{NT} that are respectively the stress correction factor and life factor for the single tooth root stress. $Y_{\delta relT}$ considers the notch while Y_{RrelT} the surface of the component. The size factor Y_X considers the dimension of the single tooth for the tooth root bending stress. It takes into account how the presence of weak points into the material, stress gradients, the presence of defect etc. [2] are influenced by the size of the gear. By consulting the normative ISO 6336 method B, for gears that have a module $m < 5$ the size factor Y_X is always equal to 1. Since this fact can lead to an oversizing of the gear, Dobler et al. [18] proposed an alternative method for computing the size factor Y_X for gears having a fine module. The size factor proposed by Dobler is reported in eqn (10)

$$Y_{XDobl} = 1 - 0.45 \cdot log\left(\frac{m_n}{5}\right) \pm 0.075. \tag{10}$$

Moreover, since in this work were performed following the STBF tests procedure and not with meshing gears, it was necessary to introduce an additional correction factor [19], [20] to take into account the slightly different loading condition. Since the objective of this work was to compute the STBF fatigue limit of the material and to evaluate its behavior under cycling loads, an inverse application of the ISO 6336 standard, including the model by Dobler et al. [18] for what concern the size factor, is applied to convert the experimental data into stresses.

2 MATERIAL AND METHODS

In order to calculate the tooth root limit value σ_{Flim}, STBF test were performed on a fine module ($m_n = 2$) gear made of 39NiCrMo3. In Table 1 are resumed the relevant properties of the gear analyzed. For the computation of Y_F, Y_S and the different correction factors, dimension needed were directly extracted from the 2D cad drawings.

Tests were performed on a SETPlab UD 04 pulsatory machine capable to apply a maximum load of 5 KN. A dedicated machine tool was appositively manufactured for correctly applying the load on the inspected tooth pair. A schematic representation of the setup used is shown in Fig. 2. As it is possible to see the gear position is constrained, during the mounting phase, by a central pin. Before starting the test, a load was applied to the gear and the pin removed. Exploiting friction, it is possible to avoid slipping of the gears before and during the test. A stress ratio $R = 0.1$ was kept during the test. According to literature [15], [21], this value is enough for avoiding unwanted movement.

Table 1: Tested gear geometry specification.

Nominal module (m_n)	2.00 (mm)
N. of teeth (z)	26 (–)
Normal pressure angle (α_n)	20 (°)
Face width (b)	20.00 (mm)
Profile shift coefficient (x)	0.30 (–)
Dedendum coefficient (h_{Fp}^*)	1.25 (–)
Addendum coefficient (h_{ap}^*)	1.00 (–)
Root radius factor (ρ_{Fp}^*)	0.38 (–)
Wildhaber (w)	5 (–)
Formal correction factor (Y_F)	2.02
Stress correction factor (Y_S)	1.90

Figure 2: Machine set up used for tests with the gear mounted in the middle.

With this specific geometry the Wildhaber $W5$ was used. In other words, $W5$ means that there are five teeth within the anvils. The Wildhaber distance is strictly connected to the geometry of the gear since it must be chosen in order to apply a load which is normal to the gear tooth surfaces. To evaluate the STBF limit the short staircase approach was used. Specifically, several tests were performed at different load levels. This method predicts to define a constant force increment ΔF. The choice of this value influences the accuracy of the final results. A low ΔF value will increase the accuracy but also the number of tests needed for reaching failure. The method works as follows. If a certain test reach fails or reach the run-out condition (withstand 5 M cycles) with a force F_i, the next test is performed at a force level defined by $F_{i+1} = F_i \pm \Delta F$ respectively. The fatigue limit at a 50% probability is computed considering the first relevant force F_1 observed that induces a failure (eqn (11)). k is a statistic factor that depends on the Run-out/Failure (RO/F) test sequence.

$$F_{FP_{STBF50\%}} = F_1 + k \cdot \Delta F \tag{11}$$

3 RESULTS

In Table 2 are resumed STBF tests performed on specimen gears. All dimensions necessary for the computation of correction factors were extrapolated directly from 2D drawings of the gear.

Table 2: Staircase results on the 2 mm module gear.

Test	$Fn_{min}[N]$	$Fn_{max}[N]$	$N\,[-]$	Status
1	−395	− 3,950	3,330,753	F
2	−385	− 3,850	$5e10^6$	RO
3	−395	− 3,950	$5e10^6$	RO
4	−405	− 4,050	417,250	F
5	−395	− 3,950	$5e10^6$	RO
6	−405	− 4,050	$5e10^6$	RO

Firstly, the force must by multiplied by the angle α_{Fen} (see Fig. 1) to compute the acting tangential force F_t. According to the staircase approach for a RO/F sequence (F-RO-RO-F-RO-RO), the value of the constant results $k = -0.296$. The application of eqn (9) leads to the calculation of the force associated to the 50% probability of failure, $F_{FP_{STBF50\%}} = 3,071.6\ N$. In Fig. 3 the result of the short staircase is plotted.

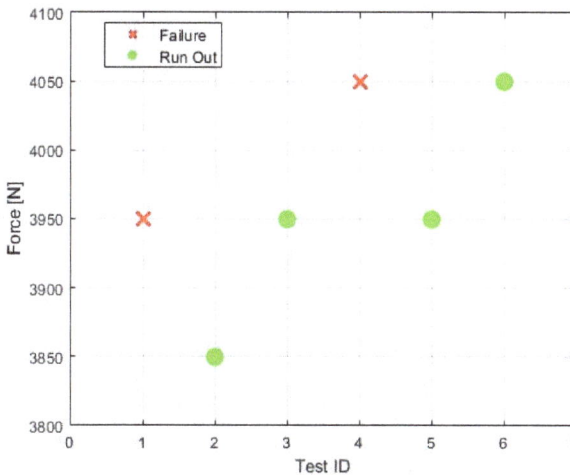

Figure 3: Staircase of the tested specimens.

According to the equations mentioned in the first section $\sigma_{F2mm} = 658.47\ MPa$. This value must now be multiplied by 0.9 to consider the fact that tests were performed on a pulsatory machine and not on meshing gears [19], [20]. With the introduction of this parameter now the effective bending stress present reduces to $\sigma_{F2mm} = 592.63\ MPa$. The correction factors necessary for the computation of σ_{FP} were chosen as follow. For our tested gear Y_{ST} is equal to 2 and Y_{NT} was equal to 1. Y_{RrelT} depends on the gear surface roughness R_z. The measured roughness R_a of the specimen was 6.3 μm, it is possible to translate the value of R_a into R_z using the DIN 4778 [22]. According to ISO 6336 the associated value to of the relative surface factor with $R_z = 38.3\ \mu m$ is $Y_{RrelT} = 0.9458$. Considering the geometry, the notch sensitivity factor computed thanks to eqn (12) was $Y_{\delta relT} = 0.9475$, were X^* is the relative stress gradient and ρ' the slip-layer thickness. (for the computation of these value please refer to [2] and [23].)

$$Y_{\delta relT} = \frac{1+\sqrt{\rho' \cdot X^*}}{1+\sqrt{\rho' \cdot X_T^*}}. \qquad (12)$$

The size factor Y_X was calculated with the new formulation proposed by Dobler et al. [18] (eqn (8)) with a resulting value of $Y_X = 1.18$ instead of the unit value proposed by the standard. Finally, the admissible root tooth bending stress computed with eqn (9) was found out at $\sigma_{Flim} = 280.44$. Moreover, thanks to the STBF tests performed it was also possible to reconstruct the characteristic S–N curve for the component. Data are plotted in Fig. 4.

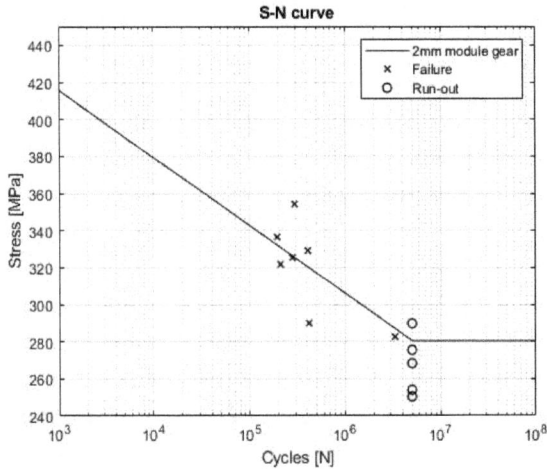

Figure 4: S–N curve of the component.

4 CONCLUSION

According to literature the admissible stress for the material 39NiCrMo3 is $\sigma_{Flim} = 280.92\ MPa$. The experimental evidence $280.44\ MPa$ confirms the increased load carrying capacity of small gears observed by Dobler et al. [18] In other words, gears made by the same 39NiCrMo3 steel show a 18% increased load carrying capacity with respect to the performances of the same materials when used on standard 5 mm module gears. The experimental tests performed in this framework, fully confirm the findings of Dobler et al.

[18]. For completeness, it should be mentioned that other formulations for the size factor are available in literature [17].

On the other hand, if standards were strictly followed, and an unitary value for the size factor is used, the present tests will lead to a σ_{Flim} equal to $330.65\ MPa$ which results about 17.7% higher than the one expected for this material. This fact highlights that there is the necessity to increase efforts to study also small gears that are and will become even more in the future, fundamental components for the mechanical industry.

REFERENCES

[1]   Punov, P., Evtimov, T., Chiriac, R., Clenci, A., Danel, Q. & Descombes, G., Progress in high performances, low emissions, and exergy recovery in internal combustion engines. *Green Energy Technol.*, **1**, pp. 995–1016, 2018.

[2]   International Organization for Standardization, ISO 6336-3. *ISO Stand.*, **3**, ISO/DTS 6336-22, 2018.

[3]   German Institute for Standardisation, DIN calculation of load capacity of cylindrical Gears—Introduction and general influence factors. 1987.

[4]   American Gear Manufactures Association, ANSI/AGMA 2001-D04 fundamental rating factors and calculation methods for involute spur and helical gear teeth. **4**, p. 66, 2004.

[5]   Gorla, C., Rosa, F., Conrado, E. & Concli, F., Bending fatigue strength of case carburized and nitrided gear steels for aeronautical applications. *Int. J. Appl. Eng. Res.*, **12**(21), 2017.

[6]   Gorla, C., Conrado, E., Rosa, F. & Concli, F., Contact and bending fatigue behaviour of austempered ductile iron gears. *Proc. Inst. Mech. Eng. Part C: J. Mech. Eng. Sci.*, **232**(6), 2018. DOI: 10.1177/0954406217695846.

[7]   Rao, S.B. & McPherson, D.R., Experimental characterization of bending fatigue strength in gear teeth. *Gear Technol.*, **20**(1), pp. 25–32, 2003.

[8]   Gorla, C., Rosa, F., Concli, F. & Albertini, H., Bending fatigue strength of innovative gear materials for wind turbines gearboxes: Effect of surface coatings. *Proceedings ASME International Mechanical Engineering Congress and Exposition (IMECE)*, **7**(Pt. A–D), 2012. DOI: 10.1115/IMECE2012-86513.

[9]   Concli, F., Austempered ductile iron (ADI) for gears: Contact and bending fatigue behavior. *Procedia Struct. Integr.*, **8**, pp. 14–23, 2018. DOI: 10.1016/j.prostr.2017.12.003.

[10] Steutzger, M., *Großeneinfluss auf die Zahnfussfestigkeit*, Frankfurt, Germany, 1997.

[11] Hertter, T., *Rechnerischer Festigkeitsnachweis der Ermüdungstragfähigkeit vergüteter und einsatzgehärteter Stirnräder*, Technische Universitaet Muenchen: München, Germany, 2003.

[12] Hoehn, B.-R., Oster, P. & Braykoff, C., Size and material influence on the tooth root, pitting, scuffing and wear load carrying capacity of fine module gears.

[13] Hoehn, B.-R., Oster, P. & Braykoff, C., Calcolo della capacità di carico degli ingranaggi con modulo piccolo. *Organi di Trasm.*, **42**, pp. 104–112, 2011.

[14] Conrado, E., Gorla, C., Davoli, P. & Boniardi, M., A comparison of bending fatigue strength of carburized and nitrided gears for industrial applications. *Eng. Fail. Anal.*, **78**, pp. 41–54, 2017. DOI: 10.1016/j.engfailanal.2017.03.006.

[15] Gorla, C., Rosa, F., Conrado, E. & Albertini, H., Bending and contact fatigue strength of innovative steels for large gears. *Proc. Inst. Mech. Eng. Part C: J. Mech. Eng. Sci.*, **228**(14), pp. 2469–2482, 2014. DOI: 10.1177/0954406213519614.

[16] Bonaiti, L., Concli, F., Gorla, C. & Rosa, F., Bending fatigue behaviour of 17-4 PH gears produced via selective laser melting. *Procedia Struct. Integr.*, **24**, pp. 764–774, 2019. DOI: 10.1016/j.prostr.2020.02.068.

[17] Concli, F., Tooth root bending strength of gears: Dimensional effect for small gears having a module below 5 mm. *Appl. Sci.*, 2021.

[18] Dobler, A., Hergesell, M., Tobie, T. & Stahl, K., Increased tooth bending strength and pitting load capacity of fine-module gears. *Gear Technol.*, **33**(7), pp. 48–53, 2016.

[19] Rettig, H., Ermittlung von zahnfußfestigkeitskennwerten auf verspannung-sprüfständen und pulsatoren-vergleich der prüfverfahren und der gewonnenen kennwerte. *Antriebstechnik*, **26**, pp. 51–55, 1987.

[20] Stahl, K., Lebensdauer statistik : Abschlussbericht, forschungsvorhaben nr. *Tech. Rep.*, (304), p. 580, 1999.

[21] Gasparini, G., Mariani, U., Gorla, C., Filippini, M. & Rosa, F., Bending fatigue tests of helicopter case carburized gears: Influence of material, design and manufacturing parameters. *Am. Gear Manuf. Assoc. - Am. Gear Manuf. Assoc. Fall Tech. Meet*, **2008**(Dec.), pp. 131–142, 2008.

[22] G. I. for Standardisation, DIN 4778, Berlin, Germany, 1990.

[23] International Organization for Standardization, ISO 6336-5. *ISO Stand.*, p. 50, 2003.

SHORT-TERM OUTDOOR EXPOSURE EFFECTS ON COTTON FABRIC ABRASION PROPERTIES

HIDEAKI KATOGI & HISAKO TSUNEKAWA
Department of Human Environmental Science, Jissen Women's University, Japan

ABSTRACT

The comfort of textile products that include cotton fibre after exposure to outdoor environments has been emphasized as a sustainable development goal. For textile products, abrasion between fabric and human skin can occur. Moreover, after abrasion, textile products present disposal difficulties. Therefore, evaluation of abrasion properties of textile products is important for comfort and reuse. This study investigated effects of short-term outdoor environments on abrasion properties of cotton fabric as a constituent material of textile products. The cotton knit fabric specimen colours were blue and red. Outdoor exposure tests were conducted based on Japanese Industrial Standard (JIS) Z 2381. Test times were 0, 2, and 4 weeks (28 October 2019–22 January 2020) for tests conducted in Tokyo (Japan). After outdoor exposure testing, abrasion resistance tests were conducted based on JIS L 1096, yielding the following conclusions. Abrasion tests indicated that the number of cycles to failure of blue cotton knit fabrics were similar to those of red cotton knit fabrics during the 4-week test period. The number of cycles to failure of all cotton knit fabrics during the 2-week test period increased compared with that of all cotton knit fabrics before outdoor exposure tests. The number of cycles to failure of all cotton knit fabrics during the 4-week test period remained almost unchanged compared to those of all cotton knit fabrics during the 2-week test period. The 2-week outdoor environment exposure affected the cotton knit fabric shrinkage because cotton fibre has a hydrophilic group of cellulose. Results suggest that structural changes of cotton knit fabric affected the fabric abrasion properties during outdoor exposure.

Keywords: abrasion property, comfortable, cotton, outdoor exposure, short-term.

1 INTRODUCTION

Recently, electronic textile (e-textile) products have been emphasized for health and sports [1]–[3]. Usually, constituent materials of fibre products are petroleum-based fibres such as nylon fibre, polyester fibre, and acrylic fibre. However, petroleum-based fibres represent disposal difficulties. Furthermore, petroleum-based fibre constituent materials of fibre products have low moisture. When petroleum-based fibre products are exposed to perspiration, bacteria can exist between the cloth and human skin. Therefore, moisture and abrasion properties of fibre products can strongly affect human comfort.

Numerous reports have described fatigue and tensile properties of natural fibres [4]–[6]. Natural fibres have good moisture and mechanical properties. For wider application, mechanical properties of green composites using natural fibre have been studied [7]–[12]. In addition, fibre products using woven and knit fabrics using natural fibres have been studied to assess their compatibility with sustainable development goals (SDGs) [13]–[20].

Tania et al. [19] reported ZnO coating effects on mechanical properties of cotton fabric. The tensile strength of cotton fabric was decreased by a ZnO coating. However, the ZnO coating increased the cotton fabric bending length.

Zhao et al. [20] reported aspects of the thermal stability of cotton fabric. Thermogravimetric analyses revealed that thermal degradation of pure cotton occurred at 298°C. The thermal conductivity of pure cotton fabric was about 0.05 W/mK.

Nevertheless, despite cotton fabric abrasion properties' importance for SDGs and wider applications, few reports have described them under outdoor exposure environments.

WIT Transactions on Engineering Sciences, Vol 133, © 2021 WIT Press
www.witpress.com, ISSN 1743-3533 (on-line)
doi:10.2495/MC210051

Therefore, this study investigated effects of short-term outdoor exposure on abrasion properties of cotton knit fabric.

2 MATERIALS AND METHODS

2.1 Materials

The 120 mm wide and 120 mm long cotton knit fabric specimens were red and blue.

2.2 Outdoor exposure testing method

The outdoor exposure test was conducted based on Japanese industrial standard (JIS) Z 2381. The 4-week-long tests (28 October 2019–22 January 2020) were conducted at Hino, Japan. A stand for outdoor exposure testing (Fig. 1) was created in our laboratory. Each of the three specimens examined was fixed at four corners. After outdoor exposure testing, the surface of each specimen was observed using ultraviolet (UV) light in a darkroom with constant temperature and humidity (20°C, 65%RH).

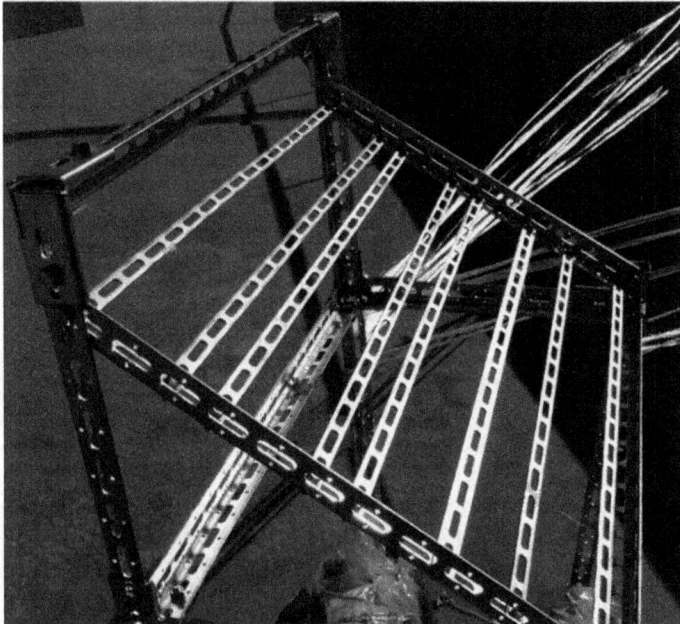

Figure 1: Stand for outdoor exposure test.

2.3 Abrasion testing method and thickness measurement

Abrasion testing of three cotton knit fabric specimens was conducted based on JIS L 1096 after outdoor exposure. The abrasion testing machine was a universal textile abrasion tester (CAT-125A; Daiei Kagaku Seiki Mfg. Co., Ltd.).

The cotton knit fabric thickness of three specimens was measured using a micrometer after outdoor exposure testing.

2.4 Ventilation resistance

After outdoor exposure, air permeability testing of three cotton knit fabric specimens was conducted based on JIS L 1096. The testing machine was an air permeability tester (KES-F8-AP1; Kato Tech Co., Ltd.).

3 RESULTS AND DISCUSSION

3.1 Surface observation and thickness measurements

Fig. 2 portrays surfaces of blue and red cotton knit fabrics before outdoor exposure.

(a)

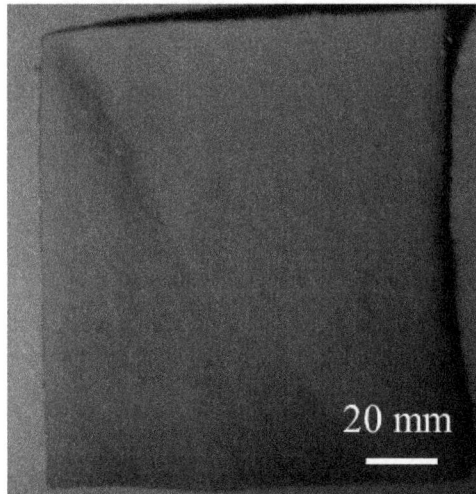

(b)

Figure 2:   Surfaces of cotton knit fabrics at blue and red colours before outdoor exposure. (a) Blue; and (b) Red.

Fig. 3 shows surfaces of blue and red cotton knit fabrics after the 4-week test period. Almost no discolouration or damage of the cotton knit fabric surface was observable after the 4-week outdoor exposure test. However, shrinkage of all cotton knit fabric specimens had occurred during the 4-week test period. Shrinkage of all cotton knit fabrics probably occurred because of moisture absorption and desorption of the cotton fibres.

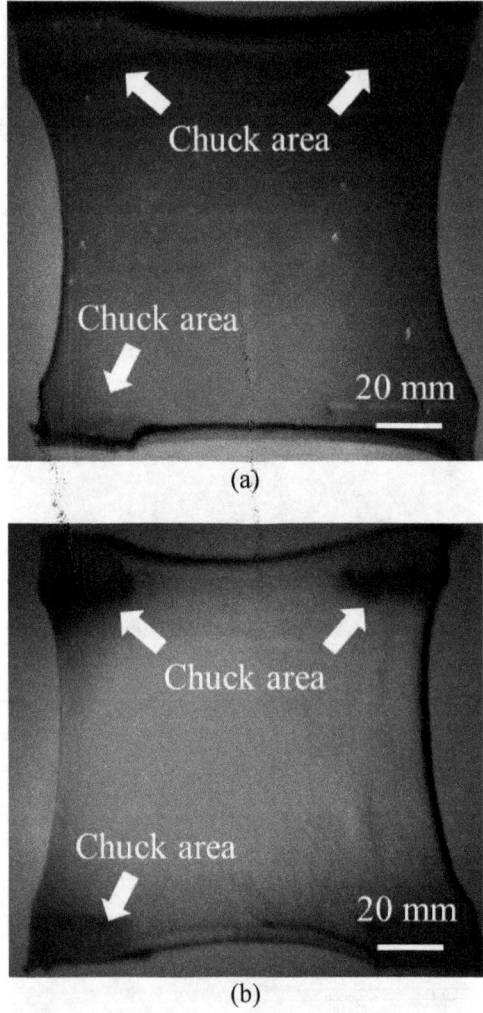

(a)

(b)

Figure 3: Surfaces of blue and red cotton knit fabrics after the 4-week test period. (a) Blue; and (b) Red.

Fig. 4 depicts blue and red cotton knit fabric thicknesses. During the test period, the red cotton knit fabric thickness increased with time. The blue cotton fabric thickness increased slightly during the 2-week test period. After the 4-week test period, the red cotton knit fabric thickness was greater than that of blue cotton knit fabric. Results show that the outdoor exposure environment slightly affected blue and red cotton knit fabric thicknesses.

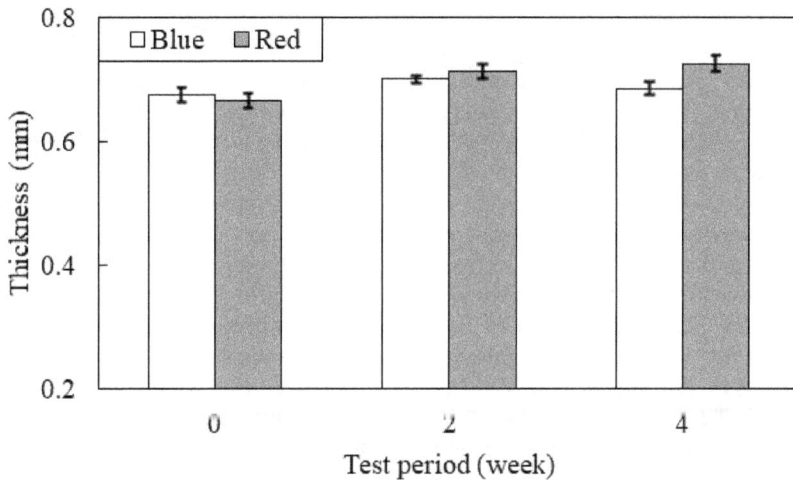

Figure 4: Cotton knit fabric thicknesses before and after outdoor exposure.

3.2 Ventilation resistance

Fig. 5 shows the ventilation resistance of blue and red cotton fabrics. Ventilation resistances of all cotton knit fabrics at the 2-week test period were greater than those of all cotton knit fabrics before outdoor exposure tests. Subsequently, the ventilation resistances of blue and red cotton knit fabrics remained almost unchanged. When outdoor exposure testing was conducted, the ventilation resistance of red cotton knit fabric became greater than that of the blue cotton knit fabric. The cotton fibre has cellulose, hemi-cellulose, lignin, and other components. The cellulose of the constituent material of cotton fibre has a hydroxyl group. Therefore, the knitting density of the cotton knit fabric probably affected the wet shrinkage of cotton knit fabric under exposure to light of these wavelengths.

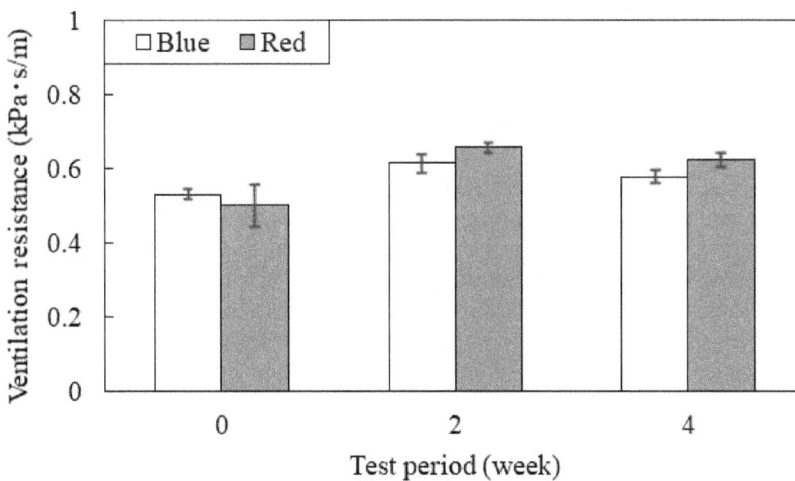

Figure 5: Ventilation resistances of cotton knit fabric before and after outdoor exposure.

3.3 Abrasion properties

Fig. 6 presents the abrasion properties of cotton knit fabrics before and after outdoor exposure testing. Before outdoor exposure testing, the numbers of cycles to failure of blue and red cotton knit fabrics were, respectively, 165 cycles and 149 cycles. The number of cycles to failure of all cotton fabrics at 2 weeks was greater than that of all cotton knit fabrics before outdoor exposure testing. Subsequently, the number of cycles to failure of all cotton fabrics remained almost unchanged with the test period duration. The cotton yarn tensile strength decreased with an increased test period during 100 h UV irradiation [21]. Generally, moisture absorption increases the tensile strength and elongation at breaking of the cotton fibre [22]. The cotton knit fabric abrasion properties were mainly affected by improvement of the fracture toughness and moisture absorbed fabric structure under outdoor exposure during the 2-week test period. To verify the abrasion mechanisms, future research must be conducted on the effects of the moisture environment on the fracture toughness and structural change of the cotton knit fabric.

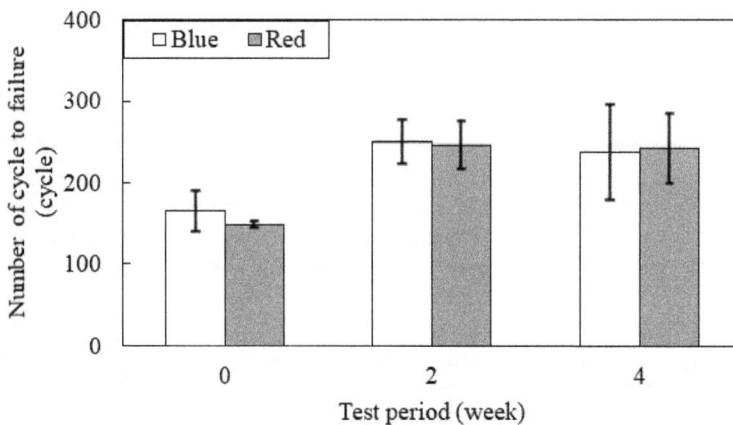

Figure 6: Abrasion properties of cotton knit fabric before and after outdoor exposure.

4 CONCLUSIONS

This study investigated the effects of short-term outdoor exposure on the abrasion properties of cotton knit fabric. The results indicate that the thickness and ventilation resistance of blue and red cotton knit fabric at colours increased slightly after a short-term outdoor exposure test. The number of cycles to failure of cotton knit fabric during outdoor exposure for the 2-week test period was greater than that of the cotton knit fabric before outdoor exposure testing. The number of cycles to failure of all cotton knit fabrics remained unchanged. The surface colour effects on the abrasion properties of cotton knit fabric were almost unchanged after short-term outdoor exposure testing. The results suggest that structural changes of the cotton knit fabric affected their abrasion properties during outdoor exposure.

REFERENCES

[1]   Grancarić, A.M. et al., Conductive polymers for smart textile applications. *Journal of Industrial Textiles*, **48**(3), pp. 612–642, 2018.

[2]   Shi, J. et al., Smart textile – integrated microelectronic systems for wearable applications. *Advanced Materials*, **32**(5), p. 1901958, 2020.

[3]   Niu, B., Yang, S., Hua, T., Tian, X. & Koo, M., Facile fabrication of highly conductive, waterproof, and washable e-textiles for wearable applications. *Nano Research*, **14**(4), pp. 1043–1052, 2021.

[4]   Katogi, H., Uematsu, K., Shimamura, Y., Tohgo, K., Fujii, T. & Takemura, K., Fatigue property and fatigue damage accumulation of jute monofilament. *Journal of the Japan Society for Composite Materials*, **41**(1), pp. 25–32, 2015 (in Japanese).

[5]   Katogi, H., Uematsu, K., Shimamura, Y., Tohgo, K., Fujii, T. & Takemura, K., Effect of cyclic frequency and time-dependent fracture on fatigue strength of jute monofilament. *Journal of the Japan Society for Composite Materials*, **41**(2), pp. 47–54, 2015 (in Japanese).

[6]   Nagasaka, T., Takemura, K., Matsumoto, K. & Katogi, H., Mechanical properties of jute fiber using the heat treatment method. *WIT Transactions on the Built Environment*, vol. 196, WIT Press: Southampton and Boston, pp. 61–68, 2020.

[7]   Katogi, H., Shimamura, Y., Tohgo, K., Fujii, T. & Takemura, K., Effect of matrix ductility on fatigue strength of unidirectional jute spun yarns impregnated with biodegradable plastics. *Advanced Composite Materials*, **27**(3), pp. 235–247, 2018.

[8]   Takagi, H., Review of functional properties of natural fiber-reinforced polymer composites: thermal insulation, biodegradation and vibration damping properties. *Advanced Composite Materials*, **28**(5), pp. 525–543, 2019.

[9]   Maruyama, S., Takagi, H. & Nakagaito, A.N., Influence of silane treatment on water absorption and mechanical properties of pla/short bamboo fiber-reinforced green composites. *WIT Transactions on the Engineering Sciences*, vol. 124, WIT Press: Southampton and Boston, pp. 101–107, 2019.

[10]   Mahdi, E. & Dean, A., The effect of filler content on the tensile behavior of polypropylene/cotton fiber and poly (vinyl chloride)/cotton fiber composites. *Materials*, **13**(3), p. 753, 2020.

[11]   Morris, R.H. et al., Woven natural fibre reinforced composite materials for medical imaging. *Materials*, **13**(7), p. 1684, 2020.

[12]   Chokshi, S., Gohil, P. & Patel, D., Experimental investigations of bamboo, cotton and viscose rayon fiber reinforced unidirectional composites. *Materials Today: Proceedings*, **28**, pp. 498–503, 2020.

[13]   Ray, R., Das, S.N., Mohapatra, A. & Das, H.C., Comprehensive characterization of a novel natural bauhinia vahlii stem fiber. *Polymer Composites*, **41**(9), pp. 3807–3816, 2020.

[14]   Madireddi, S., Aditya, K. & Hussain, S.A., Effect of combination of fabric material layers in reducing air pollution. *Materials Today: Proceedings*, **31**, pp. S197–S200, 2020.

[15]   Xu, Q., Shen, L., Duan, P., Zhang, L., Fu, F. & Liu, X., Superhydrophobic cotton fabric with excellent healability fabricated by the "grafting to" method using a diblock copolymer mist. *Chemical Engineering Journal*, **379**, p. 122401, 2020.

[16]   Mahmud, S., Pervez, N., Muhammad, A.T., Mohiuddin, K. & Liu, H.H., Multifunctional organic cotton fabric based on silver nanoparticles green synthesized from sodium alginate. *Textile Research Journal*, **90**(11–12), pp. 1224–1236, 2020.

[17]   Avadí, A., Marcin, M., Biard, Y., Renou, A., Gourlot, J.P. & Basset-Mens, C., Life cycle assessment of organic and conventional non-Bt cotton products from Mali. *The International Journal of Life Cycle Assessment*, **25**(4), pp. 678–697, 2020.

[18] Magovac, E., Jordanov, I., Grunlan, J.C. & Bischof, S., Environmentally benign phytic acid-based multilayer coating for flame retardant cotton. *Materials*, **13**(23), p. 5492, 2020.

[19] Tania, I.S. & Ali, M., Effect of the coating of zinc oxide (ZnO) nanoparticles with binder on the functional and mechanical properties of cotton fabric. *Materials Today: Proceedings*, **38**(5), pp. 2607–2611, 2020.

[20] Zhao, Z., Cai, W., Song, L., Mu, X. & Hu, Y., Comprehensive property investigation of mold inhibitor treated raw cotton and ramie fabric. *Materials*, **13**(5), p. 1105, 2020.

[21] Kikuchi, Y., Saito, M. & Kashiwagi M., Photodeterioration of silk and cotton. *Journal of Home Economics of Japan*, **38**(1), pp. 33–38, 1987 (in Japanese).

[22] Nakao, T. & Tsujim W., Restoration of hydrophilicity of field-opened cotton fibers by cyanoethylation treatment. *Sen'i Gakkaishi*, **52**(8), pp. 412–416, 1996 (in Japanese).

TOWARD UNDERSTANDING LARGE DEFLECTION BENDING OF 3D PRINTED NINJAFLEX®

LUCAS K. GALLUP, MOHAMED TRABIA, BRENDAN O'TOOLE & YOUSSEF FAHMY
Department of Mechanical Engineering, University of Nevada-Las Vegas, USA

ABSTRACT
NinjaFlex® is a thermoplastic polyurethane filament that is primarily used for 3D printing. Aside from its low cost and availability, this material is highly flexible, durable, and has a low coefficient of friction. While NinjaFlex® was introduced approximately two decades ago, a comprehensive understanding of the filament's mechanical behavior, especially in the context of the effect of 3D printing, is still lacking. The goal of this research is to gain a better understanding of the behavior of 3D printed NinjaFlex® specimens when subjected to bending loads that cause large deflection. A series of experiments were conducted to test the bending of 3D printed dog-bone specimens with different rectangular cross sections and lengths. A total of 81 specimens were tested. Test fixtures were specifically designed for this goal. Camera and image processing techniques were used to measure the deflection of the specimens. The results indicate that the experimental setup achieved its goals with specimens experiencing a maximum deflection of about 33% of the specimen length. Assessment of the experimental results showed that specimens of the same dimensions and printed under the same conditions exhibit similar deflection. Experimental data are compared to a modified form of the Euler-Bernoulli beam theory. The results of this research will lead to a better understanding of the behavior 3D-printed thermoplastic polyurethane when it undergoes large deformation in bending and to better design 3D printed components.
Keywords: NinjaFlex, polyurethane, nonlinear material, large deflection bending, 3D printing.

1 INTRODUCTION
Development of 3D printers and materials has had a major impact on engineering designs and capabilities. The thermoplastic polyurethane (TPU) NinjaFlex® has proved itself to be a highly useful printing material. It has high flexibility, high durability, and a low coefficient of friction. Relatively recent introduction of the TPU into a field mainly occupied and maintained by hobbyists means that the methodical collection and analysis of its bending behavior has not been completed. It was the goal of this paper to conduct experiments to collect the bending deflection of specimens printed with NinjaFlex® filament.

There have been earlier attempts to understand the mechanical behavior of NinjaFlex®. For example, Reppel and Weinberg [1] studied the stress strain relationship and the rupture behavior of the NinjaFlex® shell specimens under uniaxial tensile loading by varying the shell thickness of 3D printed samples. They found that Ogden material model fitted the stress strain curves of the specimens accurately. Similarly, Messimer et al. [2] studied dog-bone shaped 3D printed NinjaFlex® specimens under axial loading. The specimens had the same geometry but were printed in three different orientations. Infill density was also varied. The effects of these parameters on the stress-strain curve were considered. They found that print orientations parallel to the length of the specimens had similar stress-strain curve. It was noticed that in all cases the stress-strain curves had very small elastic regions followed by a large nonlinear portion indicating a hyperplastic behavior. It was also noticed that the ultimate strength depended on the infill density and printing orientation. Pitaru et al. [3] investigated six different filament polymer materials, including NinjaFlex®, to test their mechanical properties under axial loading when the raster angle was changed. They

WIT Transactions on Engineering Sciences, Vol 133, © 2021 WIT Press
www.witpress.com, ISSN 1743-3533 (on-line)
doi:10.2495/MC210061

found that varying raster angle of the infill of the test specimens between 0, 45, and 90 degrees had a negative effect on the mechanical properties of NinjaFlex® specimens.

While these researchers provided necessary insights into the behavior of 3D printed NinjaFlex® components, they only focused on axial loading. The aim of this work was to understand the behavior of NinjaFlex® 3D printed parts in bending. We are especially interested in characterizing large deflection of beams in bending.

The remainder of this article is organized as follows. The methodology section lays out the setup, experimentation, and processing procedures. Experimental results section describes how the specimens behaved under bending as well as a comparison with classical beam theory.

2 NOMENCLATURE

a	Distance between end of specimen and the location of the applied point load, or the load offset.
b	Width of the center section of the dog-bone specimen.
E	Young's Modulus
h	Height of the center section of the specimen.
I	Area moment of inertia of the center section of the dog-bone specimen
L	Nominal length of the center section of the dog-bone specimen.
P	Applied point load.
P_{max}	Maximum applied point load.
y_{max}	Maximum defection of the center part of the specimen.
α	Nondimensionalized applied load.

3 METHODOLOGY

To test the bending behavior of NinjaFlex®, dog-bone shaped specimens were designed. A rectangular cross-section with circular ends were 3D printed (Fig. 1(a)). The specimens were printed with a 100% infill density and the same printing orientation (Fig. 1(b)). The dimensions of the center cross-section of the specimen were varied to study the effect of the geometry on the performance of the specimens, which had one end completely fixed while loads were applied to the other end.

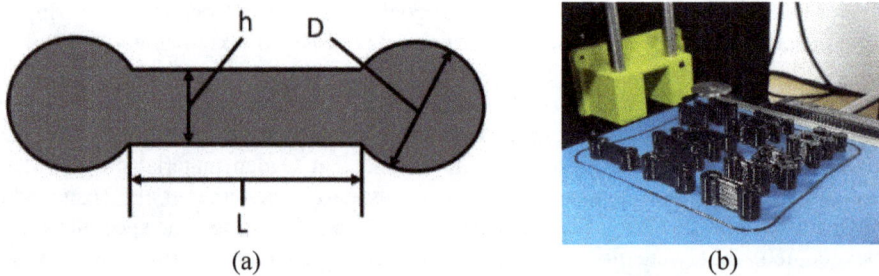

(a)   (b)

Figure 1: NinjaFlex® specimens. (a) Side view diagram; and (b) Printed specimens.

3.1 Test matrix

To understand the behavior of the material with respect to the physical dimensions, all three dimensions of the rectangular cross-section are varied between three values. Table 1 lists the values for the variable dimensions. The diameter of the cylindrical ends of the specimens were dependent on the height of the cross section, *h,* as well as the distance from the point load to the end of the center of the specimen, *a*. Variables *h, b*, and *L* were varied between *S1*, *S2*, and *S3* for each value, resulting in a total of 27 different configurations of specimens. Three samples were printed for each size for a total of 81 different specimens. It should be noted that the length in Table 1 is nominal as the rectangular region of each specimen slightly overlaps with the clamped circular end sections of the specimen.

Table 1: Specimen test matrix; all values are in millimeters.

	S1	S2	S3
h	1.8	2.7	3.6
b	8	10	12
L (Nominal)	5	10	15
D	3.85	5.7	7.7
a	18.26	17.12	16.47

The NinjaFlex® specimens were printed using a LulzBot® Mini, manufactured by LulzBot®; all test specimens were printed in the same orientation with 100% infill density. Due to some filament stringing during the printing path, some anomalies were observed. These anomalies were physically removed if they were minor. Otherwise, specimens were reprinted. Once all parts were printed and ready, they were collected and stored in an organization box. The three test samples of each size were labeled with white paint on their sides as samples A, B, and C.

3.2 Test bracket

Test brackets were designed to hold either end of a specimen by inserting it through a recess. The bracket consists of two parts. The base bracket was used to attach the test specimen to a fixed support. The end bracket had an attachment hole near the end where a load was applied by hanging weights on a nylon fishing line. The brackets were printed on a Stratasys® Fortus 250mc, manufactured by Stratasys®, using ABSplus P430 material. The specimen ID, the load incrementation, number of loads, and maximum loads are listed in Table 2.

Fig. 2(a) shows the test bracket with a NinjaFlex® specimen inside a base bracket to the left, with a slotted groove for bolt attachment to an optical table. The right side of Fig. 2(a) shows the end bracket with the hole for attaching the point load. Fig. 2(b) shows an assembled specimen at no load. The photo shows a slight bend in the specimen due to the weight of the end bracket: 3.1, 2.9, and 2.7 grams for specimen height values 1.8, 2.7, and 3.6 mm respectively.

3.3 Experimental setup

Fig. 3 shows the experimental setup with labeled components. The trials were conducted on a worktable that contains a large grid of threaded anchor points. The test bracket, located on the left side of A, was bolted down with the test side hanging over the edge of the table. The test specimen, A, was put inside the bracket, and the other end of the specimen was put in the end piece bracket. A nylon fishing line with a knot on one end was inserted through the hole in the end bracket, to the right of label A. The other end of the fishing line was tied to a swivel hook for the weights, located between labels A and B.

Table 2: Specimen ID and loading parameters for all 27 specimen configurations.

ID	Minimum load (N)	Load increment (N)	Number of loads	Maximum load (N)
1	0.0304	0.0981	10	0.9133
2	0.0304	0.0981	11	1.0114
3	0.0304	0.0981	12	1.1095
4	0.0304	0.0491	11	0.5209
5	0.0304	0.0491	12	0.5700
6	0.0304	0.0491	13	0.6190
7	0.0304	0.0491	6	0.2757
8	0.0304	0.0491	8	0.3738
9	0.0304	0.0491	10	0.4719
10	0.0284	0.0981	17	1.5980
11	0.0284	0.0981	19	1.7942
12	0.0284	0.0981	21	1.9904
13	0.0284	0.0491	23	1.1075
14	0.0284	0.0491	27	1.3037
15	0.0284	0.0491	31	1.4999
16	0.0284	0.0491	15	0.7151
17	0.0284	0.0491	18	0.8623
18	0.0284	0.0981	11	1.0094
19	0.0265	0.1962	12	2.1847
20	0.0265	0.1962	13	2.3809
21	0.0265	0.1962	13	2.3809
22	0.0265	0.1962	11	1.9885
23	0.0265	0.1962	12	2.1847
24	0.0265	0.1962	14	2.5771
25	0.0265	0.0981	16	1.4980
26	0.0265	0.0981	21	1.9885
27	0.0265	0.0981	26	2.4790

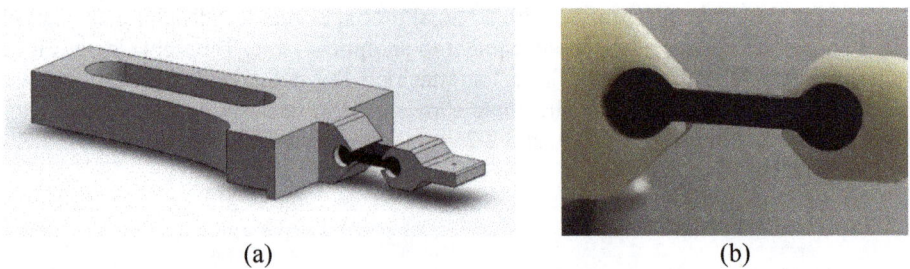

(a)

(b)

Figure 2: (a) Test bracket with NinjaFlex® specimen assembly; (b) Printed specimens.

A white board was placed behind the specimen to reduce unnecessary objects captured by the camera in the experiment. A camera was placed in the camera bracket in front of the brackets and specimen, B. A sheet metal was held above the setup with an articulating arm to reduce the glare on the top surface of the specimen cause by the ambient light in the lab.

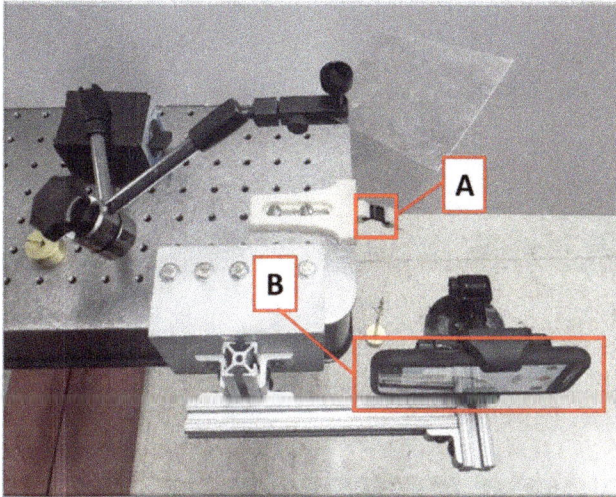

Figure 3: Labeled test setup. (A) is the test specimen and (B) is the camera.

3.4 Data collection

The data collected in the experiments were in the form of videos. The videos were captured using an iPhone 11 Pro® at 4k resolution and 60 fps. Matlab® Camera Calibration Toolbox was used to identify the intrinsic and extrinsic characteristics of the lens and thereby removing any potential distortion of the images. The camera calibration was conducted at the beginning of each experimental session.

Once calibration was completed, weights were attached to the end of the nylon line on the free end bracket in varying increments until a vertical deflection of approximately one third of the length of the center portion of the specimen was achieved. The maximum weight needed to reach this goal depended on the dimensions of the specimen.

3.5 Image processing

A custom Matlab® code was created to automatically process the images resulting from each experiment. This code used various functions of Matlab® Computer Vision Toolbox. Images were extracted from the experimental video corresponding to each load increment. Fig. 4(a) shows a typical example.

The images were undistorted based on the camera calibration results and converted to grayscale, Fig. 4(b). Next, the pixels of the images were cropped to emphasize the specimen and then converted to binary values using a custom algorithm, Fig. 5(a). The cylindrical ends of the specimen were identified using a built-in function from said Toolbox, Fig. 5(b). The radii of these circles were increased by ten pixels to account for the nonuniform contour of these circles due to pixilation. Pixels inside these two circles were converted to black, Fig. 5(c), leaving the center part of the specimen only in the images. The four edges of the remaining area were identified, Fig. 5(d). Two sets of equal number of points on the top and bottom edges were identified. Each pair of respective points were averaged to identify the corresponding point on the neutral axis, Fig. 5(e). Second-order polynomials were found to fit the top and bottom edges as well as the neutral axis, Fig. 5(f).

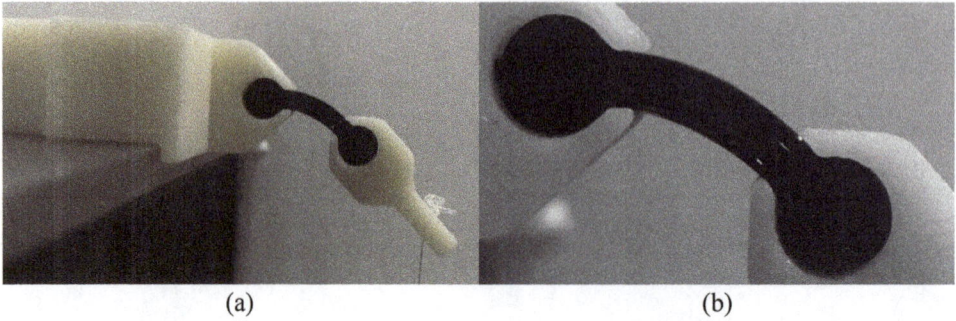

Figure 4:   (a) Raw image pulled from experiment video, specimen h = 2.7 mm, b = 8 mm, L = 10 mm nominal, Load 113 grams; and (b) Cropped and zoomed grayscale conversion of raw image.

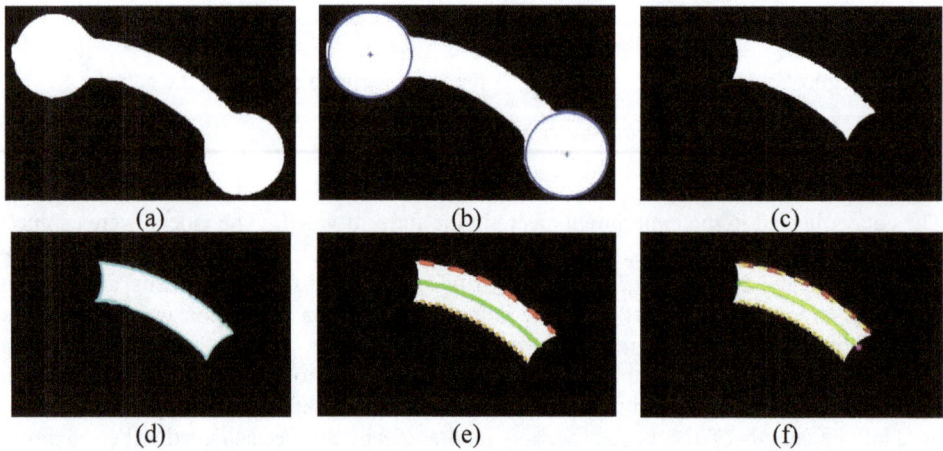

Figure 5:   Stages of processing a specimen. (a) Cropped image converted to a binary image; (b) Detection of rounded ends of specimen; (c) Removal of rounded sections; (d) Boundary detection; (e) Top edge, bottom edge, and neutral axis detection; and (f) Curve fitting of top and bottom edges and neutral axis (yellow).

The point used to measure deflection was identified as the intersection of the neutral axis fitted curve and the edge of the free end cylindrical section that was identified earlier, Fig. 6. The deflection of the three samples with identical geometry were averaged, and the corresponding standard deviation was determined.

4 EXPERIMENTAL RESULTS

Table 3 summarizes the specimen geometry and results of these experiments. The resulting bending of the specimens followed the general expected behavior of a bending beam. To help clarify the results, Fig. 7 shows the deflection of beams with the same cross-sectional area and varying lengths. Reported results are the averages of the three samples. The

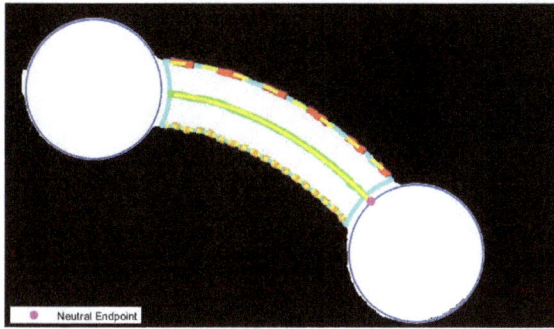

Figure 6: Final processed image, (113 grams).

increased angle at the point of load application means that the load component causing bending was gradually reduced while a load component that extends the beam was increasing. Due to the decrease in required load for desired deflection some specimens were tested with fewer loads, which can be seen in Fig. 7.

Table 3: Specimen geometry and experimental results.

ID	h (mm)	b (mm)	L (mm)	Maximum load (N)	Average x_{max}/L	Average y_{max}/L	Std. dev. x_{max}/L	Std. dev. y_{max}/L
1	1.8	8	6.925	0.9133	0.7724	0.3129	0.0171	0.0066
2	1.8	10	6.925	1.0114	0.6871	0.2803	0.0265	0.0041
3	1.8	12	6.925	1.1095	0.8870	0.2936	0.0619	0.0110
4	1.8	8	11.925	0.5209	0.7858	0.4296	0.0099	0.0207
5	1.8	10	11.925	0.5700	0.9327	0.4310	0.0058	0.0070
6	1.8	12	11.925	0.6190	0.8780	0.3827	0.1287	0.0538
7	1.8	8	16.925	0.2757	0.9145	0.4997	0.0039	0.0107
8	1.8	10	16.925	0.3738	0.8495	0.5366	0.0503	0.0021
9	1.8	12	16.925	0.4719	0.8532	0.4732	0.0158	0.0260
10	2.7	8	7.850	1.5980	0.9821	0.2480	0.0042	0.0062
11	2.7	10	7.850	1.7942	0.9726	0.2204	0.0108	0.0013
12	2.7	12	7.850	1.9904	1.0029	0.2126	0.0033	0.0044
13	2.7	8	12.850	1.1075	0.9506	0.3788	0.0097	0.0025
14	2.7	10	12.850	1.3037	0.9204	0.3433	0.0226	0.0068
15	2.7	12	12.850	1.4999	0.9398	0.3593	0.0016	0.0045
16	2.7	8	17.850	0.7151	0.8889	0.3585	0.0394	0.0211
17	2.7	10	17.850	0.8623	0.9198	0.3607	0.0062	0.0115
18	2.7	12	17.850	1.0094	0.9103	0.3598	0.0173	0.0120
19	3.6	8	8.850	2.1847	0.9594	0.2323	0.0085	0.0138
20	3.6	10	8.850	2.3809	0.8851	0.1875	0.0101	0.0067
21	3.6	12	8.850	2.3809	0.9458	0.1857	0.0037	0.0213
22	3.6	8	13.850	1.9885	0.8534	0.3853	0.0188	0.0189
23	3.6	10	13.850	2.1847	0.9448	0.3245	0.0199	0.0045
24	3.6	12	13.850	2.5771	0.9004	0.3149	0.0347	0.0015
25	3.6	8	18.850	1.4980	0.9251	0.4290	0.0188	0.0300
26	3.6	10	18.850	1.9885	0.8957	0.4554	0.0044	0.0127
27	3.6	12	18.850	2.4790	0.8801	0.4065	0.0306	0.0057

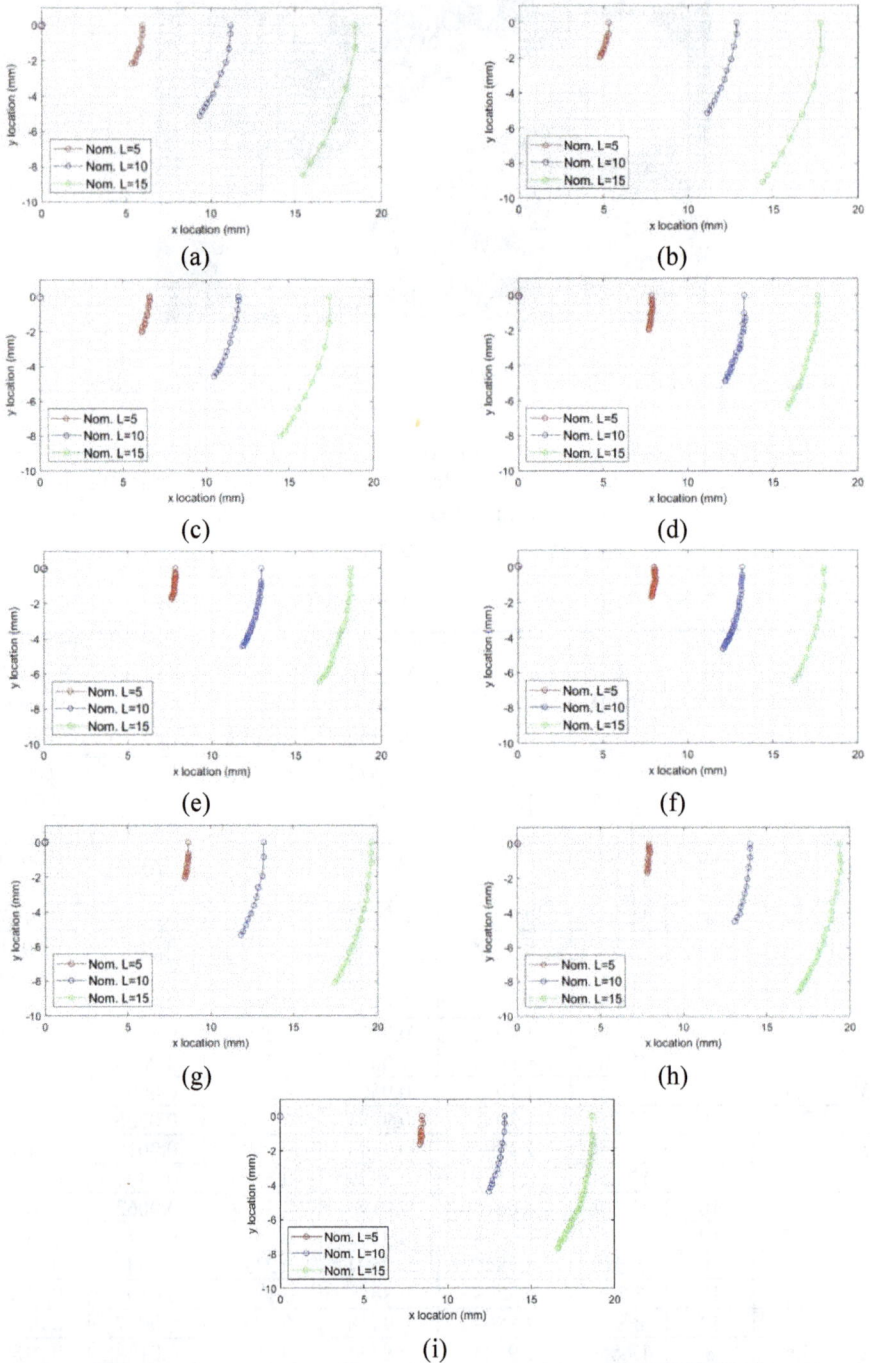

Figure 7: Deflection of specimens with the same cross-sectional area, height by width. (a) 1.8 mm by 8 mm; (b) 1.8 mm by 10 mm; (c) 1.8 mm by 12 mm; (d) 2.7 mm by 8 mm; (e) 2.7 mm by 10 mm; (f) 2.7 mm by 12 mm; (g) 3.6 mm by 8 mm; (h) 3.6 mm by 10 mm; and (i) 3.6 mm by 12 mm.

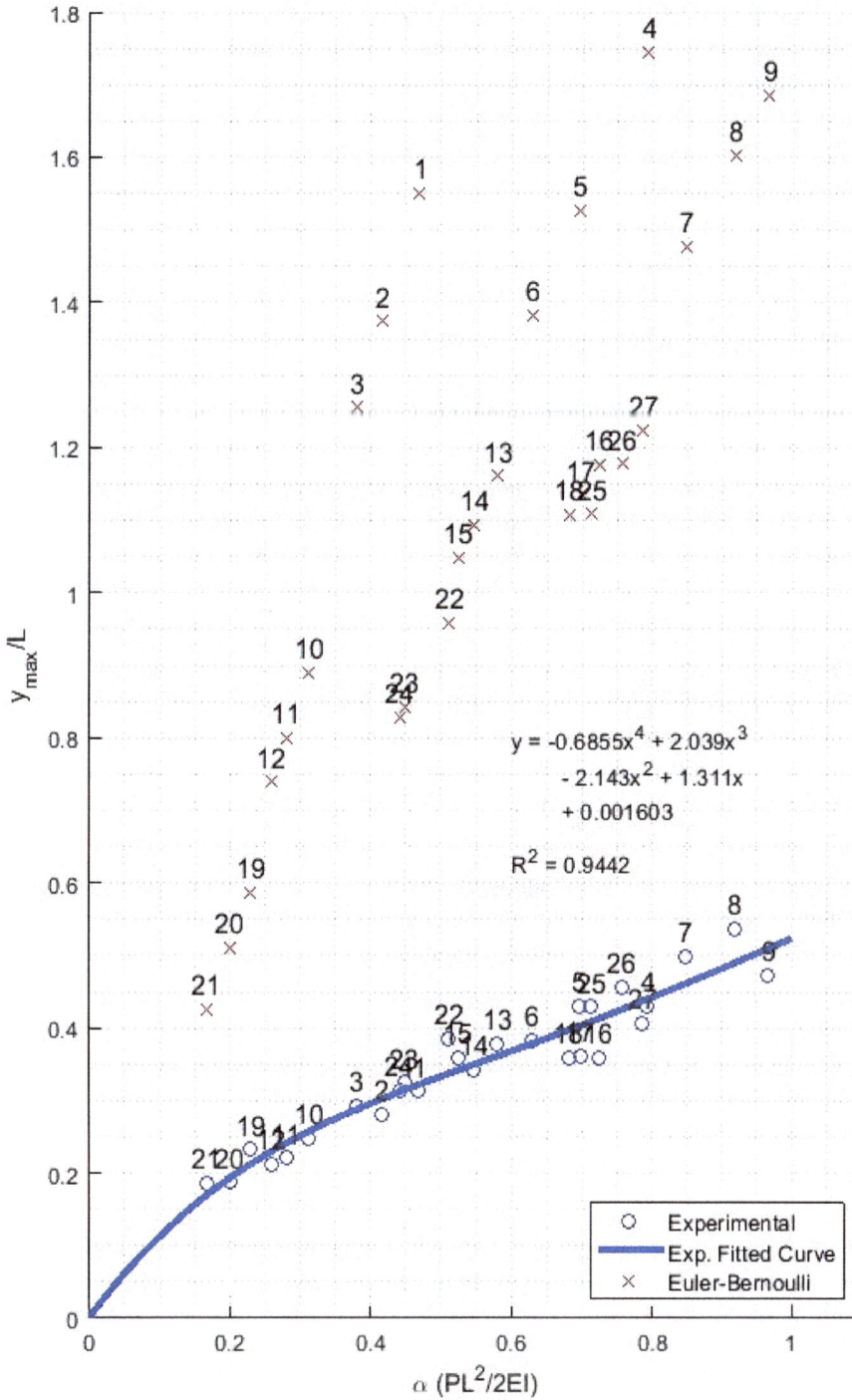

Figure 8:   Non-dimensional analysis of specimen performance with modified theoretical Euler-Bernoulli equations and fitted curve with equation.

The maximum deflections in the vertical and horizontal directions were nondimensionalized by dividing each of them by the specimen nominal length, L. The maximum applied force was nondimensionalized using a variable, α:

$$\alpha = \frac{P_{max}L^2}{2EI}.$$ (1)

Since the NinjaFlex® material can bend approximately 35% vertically with respect to the length and the load arm of the force changes with respect to the beam's slope at the endpoint, a modified version of Euler-Bernoulli will be used to approximate the bending of the specimen. It differs from the traditional theory by including deflection for both the point and moment load, eqns (2) and (3) respectively. Using superposition, a theoretical value for y_{max} was calculated by adding these two terms:

$$y_{max} = \frac{P_{max}(L+a)^3}{3EI} + \frac{P_{max}aL^2}{2EI}.$$ (2)

Fig. 8 shows the nondimensionalized maximum vertical deflection, y_{max}/L, plotted against α, the nondimensionalized applied maximum load from eqn (1). The experimental material behavior followed an approximate fourth-order polynomial trend. The R^2 value for the fitted curve was 0.9442. The red markers denote a theoretical response that should have been expected based on the modified equations. The theoretical predictions show a linear function of applied load and do not account for the nonlinear large deflections of the NinjaFlex® material. The labeled points and their corresponding dimensions can be found in Table 2.

There were several major trends in the plot. Firstly, specimens that had the same nominal length with varied depths were found grouped close together in Fig. 8, i.e., groups of three. Within these groups there was a general trend for the specimens with a larger depth to have a lower value in both the x and y axis. Similarly, specimens with the same depth and nominal length experienced a decrease in alpha value as the height increased.

A fourth order response from the fitted curve shows clearly that the material does not follow a linear pattern for large deflection. The results of the Euler-Bernoulli equations show how poorly it can model large deflection. Additionally, the Euler-Bernoulli model does not account for the hyper-elasticity of the NinjaFlex®.

5 CONCLUSION

The expanding utilization of 3D printing has encouraged the invention of novel printing filaments, such as NinjaFlex®. This material proves to be highly useful because of its strength and ability to bend. However, a full knowledge of the behavior of this material under various loading conditions is still lacking. Previous work was limited to axial loading. The goal of this research to evaluate the behavior of cantilevered 3D printed NinjaFlex® specimens when subjected to loads inducing large deformation. A test matrix was produced with 27 different specimen configurations. Three specimens were printed for each of these configurations leading to a total of 81 specimens. A combination of digital camera and custom image processing techniques were used to monitor deflection of the tip point of the specimens under various loads. The results show that specimens representing the same configuration yielded fairly close results, indicating that the behavior of NinjaFlex® is repeatable. The results of a were compared to Euler-Bernoulli theorem. The large difference in the results confirms the need for a model that can account for the hyper-elasticity of the material and the large deflection specimens experience. Future work would be focused on developing such model.

REFERENCES

[1] Reppel, T. & Weinberg, K., Experimental determination of elastic and rupture properties of printed NinjaFlex. *Technische Mechanik*, **38**(1), 2018. DOI: 10.24352/UB.OVGU-2018-010.

[2] Messimer, P., O'Toole, B., & Trabia, M., Identification of the mechanical characteristics of 3D printed NinjaFlex®. *Proceeding of the ASME 2019 International Mechanical Engineering Congress and Exposition. vol. 9: Mechanics of Solids, Structures, and Fluids*, Salt Lake City, Utah, USA. 11–14 Nov. 2019. V009T11A004. ASME. DOI: 10.1115/IMECE2019-11674.

[3] Pitaru, et al., Investigating commercial filaments for 3D printing of stiff and elastic constructs with ligament-like mechanics. *Micromachines*, **11**, 2020. DOI: 10.3390/mi11090846.

NUMERICAL AND MATERIAL MODELLING FOR THE DEVELOPMENT OF A NEW DEVICE FOR THE TREATMENT OF INDURATIO PENIS PLASTICA

PAVEL DRLIK, SARKA PESKOVA, VLADIMIR KRISTEK, MIROSLAV PETRTYL,
MARTIN VALEK, JIRI LITOS, PETR KONVALINKA & RADEK SEDLACEK
Czech Technical University in Prague, Czech Republic

ABSTRACT
The aim of the work is numerical and material modelling for the development of a new device for the treatment of Induratio penis plastica (IPP) also known as Morbus Peyronie, where it is necessary to specify the parameters of the new device for its production in cooperation on a project of the Czech Technical University in Prague and the company Medipo-ZT,s.r.o. Shock wave method was applied for patients. Evaluation and study on models created from cement mixtures with different degrees of hardness was also included, because of waiting for a real sample taken after the plaque operation. The approximate material properties of collagen are taken into account in the models. An integral part of the work is the evaluation of individual important indicators, such as plaque size, penile angulation, the international questionnaire of erectile dysfunction and the age of patients before and after treatment. At the same time, the intention is to develop a relationship between the individual indicators. Applied shock wave therapy was used for the solution, which is used in urology in the treatment of urinary stones. In physiotherapy and orthopaedics, it is used in the treatment of calcification and non-calcification diseases, such as plantar fasciitis, tennis or golf elbow, heel spur and humeroscapular periarthritis. The project is further developing software to help doctors outline the success and scope of treatment for a particular patient of a given age.
Keywords: Morbus Peyronie, Peyronie's disease, shock wave, dorsal angulation, size plate, numerical model, material model, new device, developed software.

1 INTRODUCTION
Induratio penis plastica (IPP) was described in 1561 by Fallopius and Vesalius as fibrous cavernousitis. This disease was also dealt with more thoroughly by the personal physician of King Louis XIV of France, Francois de la Peyronie, who conducted a study on plastic induration of the penis and recommended therapeutic approaches in the treatment of this disease (1743) [1]. Since then, the term induratio penis plastica or Morbus Peyronie has been used. In the past, a number of physicians and scholars (Bishop of Theodoric, Bologna – 13th century) dealt with the treatment of this deformity. Older authors considered induratio penis plastica to be a rare disease with a high tendency to regression, but recent studies report only exceptionally spontaneous disappearance (up to 2–3%)

Epidemiological data are limited, the prevalence of IPP has been rising in recent decades. Today it is in the range of 6–9% of men in Europe [1], in the USA it is slightly higher and in the range of 7–11%. Nevertheless, a number of studies suggest that the disease is often overlooked and underdiagnosed and its prevalence is even higher. The prevalence is mainly increased by the current incidence of erectile dysfunction, diabetes, high blood pressure, hyperlipoproteinemia, smoking and alcohol consumption. A typical patient is a man between 55 and 60 years old. IPP is a disease that locally affects the tunica albuginea, most commonly on the dorsum and flank of corpus cavernosum.

The aetiology of this disease is not fully understood. Recently, a disorder of wound healing (so-called microvascular injuries and traumas caused during sexual intercourse) is being considered. Prolonged healing and abacterial inflammation lead to the formation of a

WIT Transactions on Engineering Sciences, Vol 133, © 2021 WIT Press
www.witpress.com, ISSN 1743-3533 (on-line)
doi:10.2495/MC210071

scarred fibrotic plaque, which loses the elasticity of the tunic of the albuginea [2], [3]. As part of the healing, calcium salts also begin to fall into this plate, which further potentiates the rigidity of the plaque. The fact that this disease is one of the autoimmune diseases with impaired healing is also indicated by the high incidence of patients who are also affected by other autoimmune diseases of connective tissue (Dupuytren's contracture). The prevalence is 9–39%.

The disease manifests itself in several ways. Some patients report a sudden onset of penile angulation without ever having a painful erection. In others, a painful erection first appears, which is accompanied by a curvature of the penis after a few months. Most patients complain of penile shortening. From the acute phase, Peyronie's disease passes to a resting stage, which is characterized by the permanent presence of deformity. Pain persists at rest in 35–45% [4]. Deformation is expressed by the curvature and shortening of the penis during erection. The formation of a solid scar (plaque) results in the development of penile dysfunctions. In more than 50% of men, the penis is deformed during erection and thus difficulties during sexual intercourse, which negatively affects the quality of life not only of affected men, but also of their partners.

We divide the therapy of induratio penis plastica into conservative and surgical. We prefer conservative therapy in patients early after the transition from the acute to the chronic phase. In the acute stage, we recommend antiphlogistic and analgesic treatment with non-steroidal analgesics. We divide conservative treatment into oral and injectable directly into the plate.

Conservative therapy involves the oral administration of vitamin E, colchicine, L-arginine, injection of steroids, verapamil or interferon directly into the plaque. The most effective methods include the application of shock waves (SWT) [4] and the injection of Clostridial collagenase (CCH). Surgical treatment is recommended after failure of conservative therapy (surgery according to Nesbit 1965, according to Yaschia 1993) [5]. Today, only a minimum of patients undergo surgical treatment. The reason is the success of conservative therapy (especially SWT and CCH), the risk of recurrence and possible postoperative complications, including erectile dysfunction. Often, surgery is also associated with shortening of the penis during erection, as a result of surgical techniques.

2 THE DEVICE DEVELOPMENT PROCESS

The aim of the present project is industrial research on a new medical device for the treatment of Morbus Peyronie patient using new diagnostic techniques in active collaboration with a potential customer (the Central Military Hospital in Prague, hereinafter UVN).The new facility will involve a wider range of generations effect of shock waves through the applicator applied to the tissue being treated and will include patient comfort (seat attachment etc.). The new facility will be equipped with software, which will help doctors determine the intensity of the shock energy, including the number of shock waves, as well as the length of treatment. The software is based on a database of 150 patients, where the treatment is evaluated depending on the size of the plaque, the curvature of the penis, the intensity of treatment and its success.

For the new device facility, it was necessary to address several construction details, in particular:

1. Functional sample control unit
2. Keyboard
3. Power part of the shock wave generator
4. Generator control unit

5. Shock wave applicator with aiming and fixing device
6. Adjustable spark gap with motor and motor mechanism for spark gap replacement
7. Applicator positioning motors
8. Water treatment plant
9. Development of an ellipsoid with marking of foci
10. Patient's chair
11. Source part
12. Fixation and orientation of the affected area
13. Software.

Fig. 1 shows shock wave applicator-ellipsoid with marked foci; Focus F1-spark gap in the aquatic environment; and Focus F2-therapeutic focus approx. Ø9 × 38 mm. Fig. 2 shows pressure profile in focus.

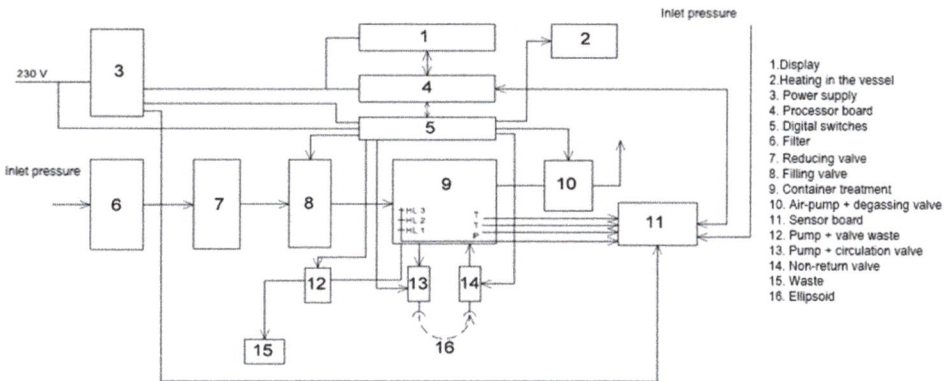

Figure 1: Shock wave applicator-ellipsoid with marked foci.

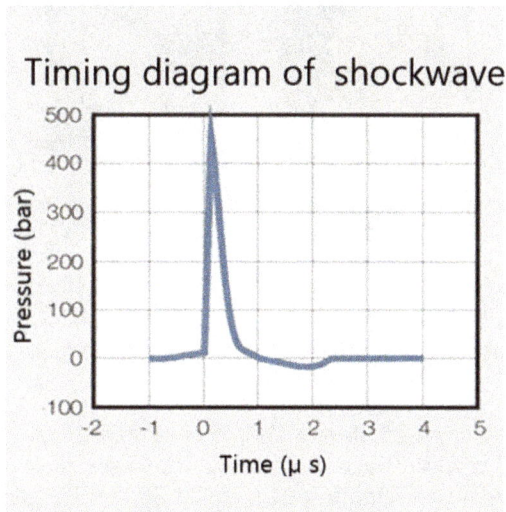

Figure 2: Pressure profile in focus.

3 SPECIFICATION OF MATERIAL MODELS

The project managed to obtain a database of approximately 150 patients suffering from the disease and which underwent shock wave therapy on devices that did not serve directly to treat the disease, so the success rate of treatment was about 75%. The database contained data on plaque size, patient age, penile curvature, the number of sessions and the total amount of energy used in the operation, and the database contained data on whether the patient was cured or not. Material models served primarily as a feedback analysis to create a new device, i.e. focus adjustment, attachment, application of energy, as current devices do not allow this. The current existing devices are primarily manufactured for other purposes and thus create inaccuracy in targeting, including very uncomfortable patient, who often has to assist doctors with attachment throughout the procedure.

In the context of this project, 43 models of penises with plaques were produced, the material was designed according to the experience of surgeons from Military University Hospital Prague (UVN). The models were tested on the Litotryptor MEDILIT device, a standard loading procedure of 3,000 cycles. The models were manufactured to test the accuracy of the shock wave impact and alignment, including the proposed mount and foci of the new device. The outputs were tested on a Phenom XL microscope, which is owned by CTU in Prague, and were compared with unloaded samples.

The material samples are composed of collagen bags with cement plaque models from C25/30 to C35/45. Fig. 3(a) shows samples of the cement plaques produced from collagen models (Fig. 3(b)) and detail of model focus to focus (Fig. 3(c)) including partial destruction after surgery (Fig. 3(d)). In the experiment, it was evident that the accuracy of focusing on the focus of the plaque and the attachment of the penis model plays an extremely important role in making the procedure more successful. That is why the team members have already designed a laser focus and an inflatable penis fixation element.

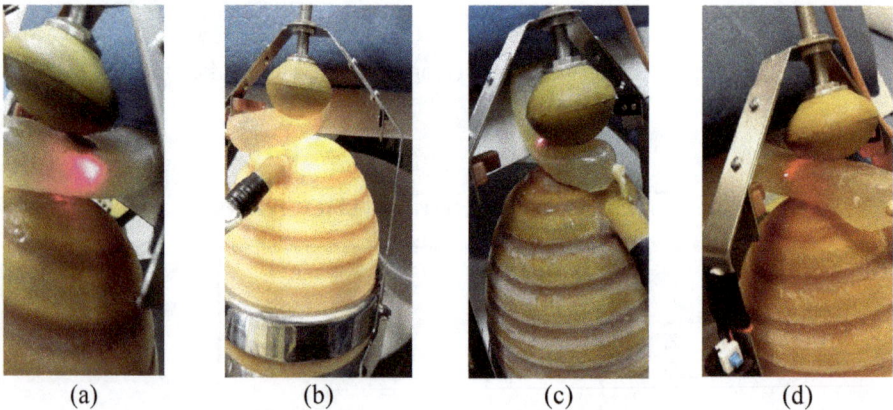

(a) (b) (c) (d)

Figure 3: Parts (a)–(d) show a laser aiming test and an inflatable pad for measurement on an artificial penis sample with plaque according to the actual dimensions.

Twelve samples of class C25/30 and C35/45 were examined with a microscope. All tested samples showed cracks after loading with shock waves; in some they were visible to the eye. The results from two samples of Class C25/30 are illustrated in Fig. 4. Samples was subjected to shock waves. For accuracy, three places in the micro level were focused

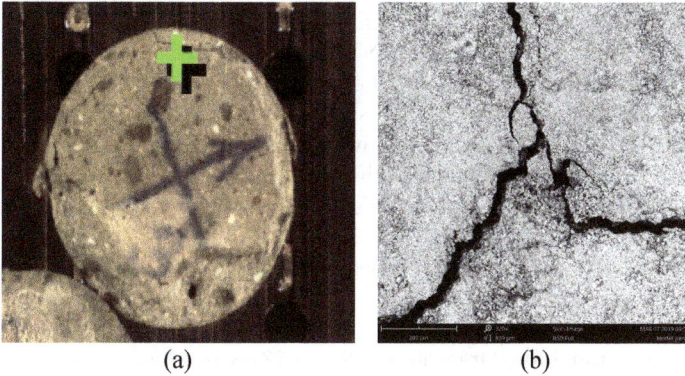

(a)   (b)

Figure 4:   Parts (a) and (b) of the sample show places in the micro level were focused under a microscope which were subjected to shock wave.

under a microscope; in sample number one there is a noticeable decay, with sharp cracks across the whole sample, which is caused by shock waves. However, the samples served rather to accurately determine and specify the focus of the instrument.

A numerical model was established for the treatment of patients and the dosing of the amount of energy, which was based on the actual material values measured by the pathologist and mechanical tests of the plaque removed during surgery. In this respect, the data used in the article are unique, because due to the risks of the operation, the number of procedures is negligible. The results from the histology of plaque, namely the population of spindle fibroblasts in the phase of chaotic orientations and in the phases of oriented fibroblasts in the directions of main stresses/strains and bone beam structure lined with fibroblasts, copying the boundaries between beam and fibrous scar are illustrated in Fig. 5. From pathological examination with the cooperation of bioengineers which was carried out by Prague's Institute of Pathology" it was determined that the modulus of elasticity of the plasma is between 15–20 MPa and the Poisson's ratio is 0.4. Further mechanical tests were performed in a specialized laboratory of CVUT in Prague, which are described below.

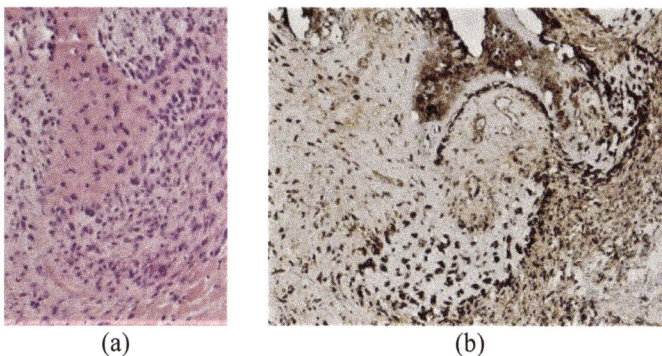

(a)   (b)

Figure 5:   Parts (a) and (b) show the results from the histology of plaque, namely the population of spindle fibroblasts in the phase of chaotic orientations and in the phases of oriented fibroblasts in the directions of main stresses/strains and bone.

4 ASSESSING OF MATERIAL PROPERTIES FROM EXPERIMENTAL ACTIVITIES

The aim of the tests was to evaluate the mechanical properties of the supplied biological plaque samples during pressure tests. Two samples were delivered for the implementation of experiments. The evaluation parameters were the elastic stress gradient (direction of the tangent in the relief phase) and the absorbed energy up to 50% deformation. A methodology developed, verified and validated in the mechanical testing laboratory was used for testing.

The tested plaque samples were supplied from the Institute of Pathology of the 1st Faculty of Medicine, Charles University and the General Hospital.

The MTS Mini Bionix 858.02 system (metrological designation PM 00, accuracy class 1, see Fig. 6) with a fitted force transducer (PM 00/18) with a range of $0 \div 500$ N was used for testing. Special jigs for pressure tests were also used (see Fig. 7). A program created in FlexTest GT software called: Pressure modul.000 was used to control the test.

Figure 6: Testing system MTS Figure 7: Image close-up shot of
Mini Bionix.   a sample save.

The MAHR 16EX digital caliper and the KINEX caliper were used to measure the samples. The environmental conditions were recorded by a digital thermometer-hygrometer COMET (PM 07). All used meters are registered in the quality management system of the laboratory and are metrologically linked to standards. Calibrations take place at regular intervals at accredited calibration laboratories.

4.1 Experimental methodology

The methodology of the experiment was based on the determination of the elastic stress gradient in the relief phase and the absorbed energy. For this type of material, the elastic stress gradient can be considered comparable to the Young's modulus. The principle of the test was loading the test specimen inserted between two metal plates at a constant feed rate to F1, unloading to the level F0 and repeating the loading to the forces F2, F3, F4, F5 and always subsequent unloading to F0. The loading speed was set at 2.0 mm/min and was the same for the unloading speed. The sampling frequency of the quantities (time, force, displacement) of the experimental test system MTS Mini Bionix 858.02 was 30 Hz.

4.2 Measured values

The samples were approximately cuboid in shape. Table 1 below summarizes the determined dimensions of the individual samples together with the selected levels of loading forces. Sample No. 1 was only for calibration purposes and was made of a different material, so it is not listed below.

Table 1: Dimensions of samples.

Number of samples	Dimensions of samples			Levels of loading forces				
	Width (mm)	Length (mm)	Height (mm)	F1 (N)	F2 (N)	F3 (N)	F4 (N)	F5 (N)
2	2.53	5.50	1.85	15	30	45	60	75
3	3.27	5.62	2.77	15	30	45	60	100

4.3 Evaluation

A contractual diagram (stress-strain curve) was prepared for each sample loaded by the pressure test according to the above methodology. By evaluating this diagram, the tangent guidelines in each relief-load cycle and the absorbed energy were determined. The stress result for sample numbers 2 and 3 are shown below in Figs 8 and 9.

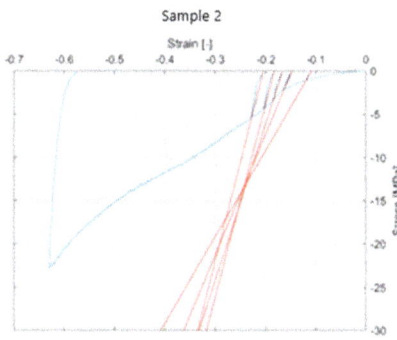

Figure 8: Diagram of sample 2 for stress–strain curve.

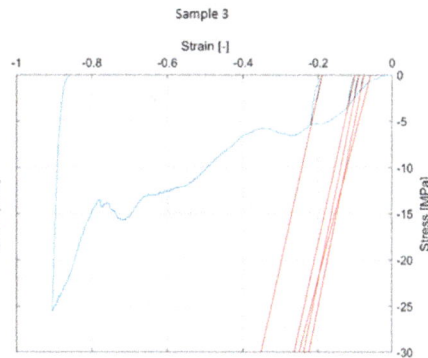

Figure 9: Diagram of sample 2 for stress–strain curve.

The points through which the secant (line) is intersected are shown in the graph as the maximum value of the stress in the relief phase and as the first point from which the relief phase began. The linear line in the form $y = a \cdot x + b$ is led by the points found in this way, where the parameter a corresponds to the elastic stress gradient.

The absorbed energy is proportional to the area under the stress–strain curve up to 50% of the strain (see Figs 10 and 11).

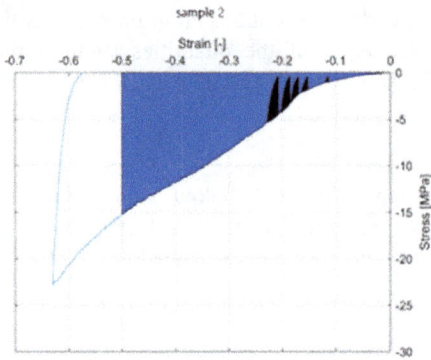

Figure 10: Diagram of energy – sample 2.

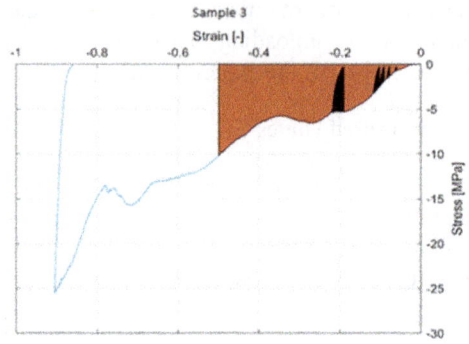

Figure 11: Diagram of energy – sample 3.

4.4 Results

The summary results of the elastic stress gradient and the absorbed energy for both samples are given in the following Table 2 and resulting values of absorbed energy are given in the Table 3.

Table 2: Elastic stress gradient.

Elastic stress gradient (MPa)		
Cycle	Sample 2	Sample 3
F1 – F0	100	155
F2 – F0	140	202
F3 – F0	181	201
F4 – F0	224	183
F5 – F0	231	183
Arithmetic mean	175	185
SD	55	19
Median	181	183
LQ (lower values)	140	183
UQ (upper values)	224	201

Table 3: Resulting values of absorbed energy.

Results of absorbed energy (MJ.m^{-3})	
Sample 2	Sample. 3
3.24	2.58

5 NUMERICAL MODELING OF PENIS DEFORMATION IN INDURATIO PENIS PLASTICA

For numerical modelling, program ATENA (Advanced Tool for Engineering Nonlinear Analysis) [6] was used. It's based on finite element method and fracture mechanics, which is very suitable for solving this problem. Several models were created in longitudinal and transverse section of the penis (Fig. 12). The supports correspond to the designed device for attaching the penis and only prevent movement in the normal direction. The load is continuous, perpendicular to the horizontal projection of the element.

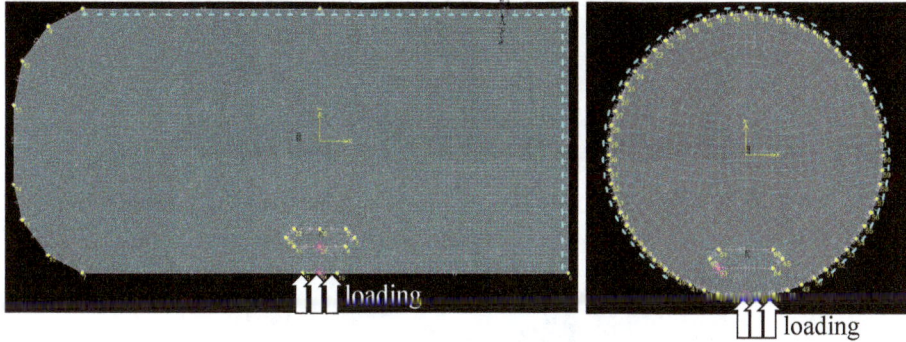

Figure 12: Models in longitudinal and transverse section of the penis.

The penis is not of a homogeneous material, but the material characteristics of its soft parts are quite similar. Therefore, only two types of material are used in the model. One homogenized for the penis where Young's modulus of elasticity was 30 kPa and Poisson's number was 0.48. The second was material of penile plaque. Here the material characteristics obtained from pathological examination was used. Young's modulus of elasticity was 80 kPa and Poisson's number was 0.46.

At first the results from model of the entire cross section of the penis is introduced. The triangular elements were used in this model (1716 elements, 905 nodes). It can be seen on them that only a small area around the plaque is affected by the load (see Figs 13 and 14). Therefore, only the section of this area is further modelled, which allows a more detailed network of elements. In Fig. 15, a cut-out from the longitudinal section of the model is depicted, with the variation of normal stress in horizontal direction σ_x. In this model were used quadrilateral elements (10,490 elements, 10,759 nodes). As well as in the cross-section model, only the variation of stress in a small area around plague is of interest. There were used 5,153 triangular elements with 2,842 nodes in this model. The crack development is depicted in Figs 16–18 together with normal stress in horizontal direction σ_x.

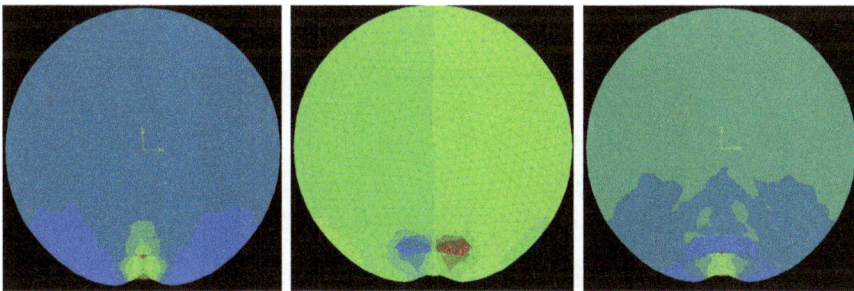

Figure 13: The variations of normal stress in horizontal direction σ_x, tangential stress τ_{xy} and normal stress in vertical direction σ_y.

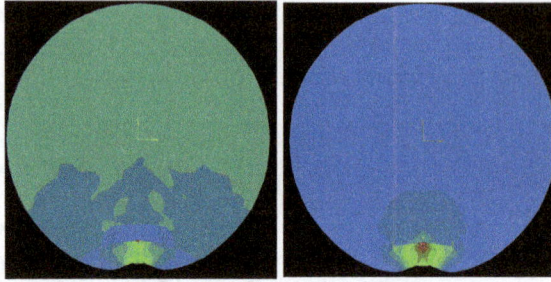

Figure 14: The variations of main stress σ_{max}, (left) and σ_{min} (right).

Figure 15: Cut-out from model of longitudinal section, with the variation of normal stress in horizontal direction σ_x.

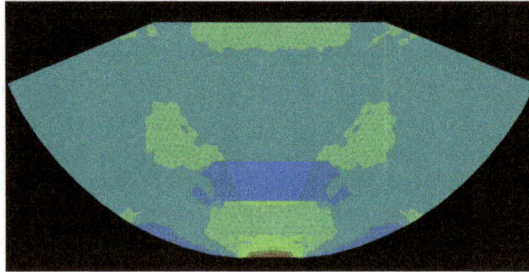

Figure 16: Model of section of area. Variation of normal stress σ_x, the state just before the formation of cracks.

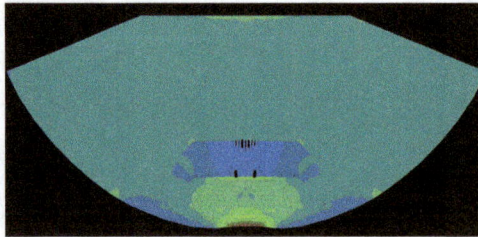

Figure 17: Initial stages of crack development.

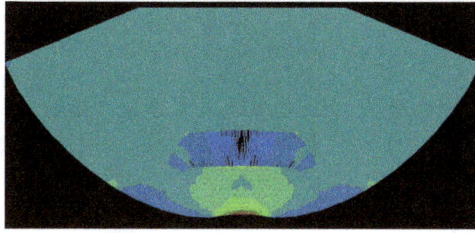

Figure 18: The final stage of crack development.

CONCLUSION

The aim of the project was research on a new medical therapeutic agent for the treatment of Induratio Penis Plastica (i.e. Peyronie's disease, IPP), using new diagnostic and therapeutic methods, with the application of shock waves. In connection with the stated goals, the following were processed: histological analyses of plaques, numerical analyses and experimental verification of elasticity/stiffness of biomaterials and their correlation with the computational modulus of elasticity of plaques. The numerical model contributed to the demonstration of real therapeutic effects (such as tests on patients), which took place at the Central Military Hospital in Prague. The numerical model and experimental analyses proved the validity of the relevant biomechanical axiom: The angle of deviation of the penis axis in IPP is dominantly dependent (i.e. a function) of plaque stiffness, its 3D size and the hydraulic pressure of blood flow during erection.

Research proved that the early phase is histologically characterized by undirected, structurally disordered, type 1 collagen fibers. Early plaques are 3 to 5 mm in 3D size. In the central areas, they contain small districts of beginning ossification. In the early phase, there are no visually visible deviations of the penis median angle. However, mature plaques already contain longitudinally oriented fibroblasts and collagen fibers in the directions of dominant major tensile stresses. In the central parts they have fractal or compatible osteogenic islands (microlayers). The matured plaques are 6 to 15 mm in 3D size. In the early phase, the stiffness of the plaques is relatively small, the angular deviations of the penis median are not visually visible. In the mature phase, the plaque stiffness gradient clearly increases. The greater the dynamic modulus of elasticity of the plaques, the greater the deviation of the penis median angle. In clinical biomechanics, the more precisely the dynamic modulus of elasticity of plaques is verified, the more precisely the angle of the penis median is defined, in the mature phase of plaque genesis and subsequently the more effectively the therapeutic procedure is chosen. From the measurement of dynamic modulus of elasticity in modeled penises, it is clear that in fully inorganic (artificial) plaques, the dynamic modulus of elasticity reaches values in the order of GPa units and also the largest deviation of angles. Ultrasound measurements of dynamic modulus of elasticity in the hospital on patients have shown that the sizes of dynamic modulus of elasticity of plaques at the interface of their early and maturation phase of development have significantly lower values of dynamic modulus, which reach numerical sizes in tens of MPa. The higher the hydraulic pressure of the blood, the greater the hydraulic strengthening of the penis and the greater the gradient of the deviation of the deviation of the penis center angle. Direct ethical pressure measurements directly in the penis during erection could not be performed on patients for ethical reasons. However, direct measurement on the plates of many patients by ultrasound, sufficiently takes into account the dynamics of hydraulic loads and is sufficient for stress analysis of connective tissues and their transformation by FEM.

Currently, the work on the project is gradually being completed. Fig. 19 shows an example of a condensed function, enabling efficient control – software calibration. During the execution of the project, the above-mentioned author's team obtained a patent for the Czech Republic, which will be further extended for the USA and Germany. Next year, extensive clinical verifications will be performed on new devices, enabling effective therapy. We expect the therapeutic efficacy of IPP to increase by up to 85–90%.

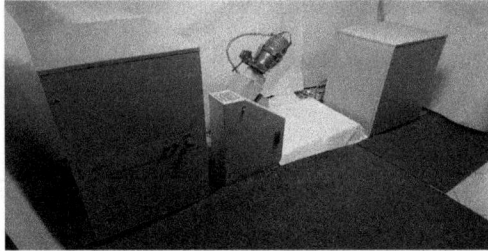

Figure 19: Functional sample of a new device for the treatment of Peyronie's disease.

ACKNOWLEDGEMENT
This study was supported by grant number TH03010470 by agency TACR.

REFERENCES
[1]   Muller, A. et al., The impact of shock wave therapy at varied energy and dose levels on functional and structural changes in erectile tissue. *European Urology*, **53**, pp. 635–643, 2008.
[2]   La Peyronie, F.G., Sur quelques obstacles qui s'opposent á l'ejaculation naturele de la semence. *Mémoires de L'Académie Royale de Médecine*, **1**, pp. 337–342, 1743.
[3]   Van de Water, L., Mechanisms by which fibrin and fibronectin appear in healing wounds: Implications for Peyronie's disease. *Journal of Urology*, **157**, pp. 306–310, 1997.
[4]   Devine, C.J., Introduction to Peyronie's disease. *Journal of Urology*, **157**, pp. 272–275, 1997.
[5]   Williams, G. & Green, N.A., The non-surgical treatment of Peyronie's disease. *British Journal of Urology*, **52**, pp. 392–395, 1980.
[6]   Software ATENA 2D Verse 5.4.1.0, Cervenka Consulting: Manual users, Department of Mechanics.

SECTION 2
RECYCLED AND
RECLAIMED MATERIALS

TREATMENT AND REUSE OF ASH FROM MUNICIPAL SOLID WASTE INCINERATION

CINZIA SALZANO, MARCO DE PERTIS, GIOVANNI PERILLO & RAFFAELE CIOFFI
University of Naples "Parthenope", Italy

ABSTRACT
Incineration is considered one of the most convenient treatment of urban solid waste (MSW) as it allows a significant volume reduction and an energy enhancement of the waste itself. However, it cannot be considered a final treatment solution because of the formation of solid residues, mainly composed of two groups of ash, bottom ash (BA) and fly ash (FA). Their characteristics depend on the type of incoming waste and on the combustion methods. BA are considered non-hazardous waste, while FA, due to their high content of heavy metals, alkali chlorides and soluble metal salts, have the characteristics of hazardous waste. Among the various recovery possibilities is the use of FA for the production of artificial lightweight aggregates (LWA), used for the production of lightweight concrete (LWC). This article aims to highlight how the use of FA granules as aggregates in LWC can give good results in terms of compressive strength, rupture and elastic modulus. In fact, the particle size distribution and chemical composition of the FA, as well as the generally spherical shape and low cost, make this type of ash an ideal material for this use.
Keywords: incineration ash, lightweight aggregates, granulation, lightweight concretes.

1 INTRODUCTION

The use of artificial aggregates obtained from waste materials and by-products, as an alternative to natural aggregates, has aroused considerable research interest. Many industrial wastes, including soil waste, fly ash from municipal solid waste incineration (MSWI-FA), blast furnace slag (GGBFS) and marble sludge (MS), can be applied to make lightweight aggregates [1].

Municipal solid waste incineration ash (MSWI-FA), for example, which contains heavy metals, chlorides and sulphates potentially harmful to cement materials and human health, requires pre-treatment before it can be safely disposed of in landfills or recycled in the construction sector [2].

Extensive research was also conducted to pre-treat different types of waste to be used as raw materials for the production of artificial aggregates. Artificial aggregates can be produced by two types of processes: cold granulation and high temperature sintering. The cold granulation process has recently received quite significant attention in the literature which is rich in applications where many waste materials have been shown to have the potential to be used as feedstock: fly ash from combustors and municipal solid waste incinerators, metallurgical slag, furnace dust, sediment, shredding waste. Their suitability for the production of artificial aggregates is undoubtedly worthy of consideration. It is known that waste treatment is largely based on cement-based stabilization/solidification processes, which allow for safer disposal and/or recovery of materials for the manufacture of building materials [3].

These systems have shown promising results in terms of physical, mechanical and resistance properties and the possibility of synthesis starting from industrial waste, favoring economic benefits both in the field of waste cycle management, such as the reduction of material to be disposed of in landfills, and concrete ecological and energy advantages.

Therefore, in this paper, a study on the recycling of fly ash (FA) by the cold granulation process is proposed.

WIT Transactions on Engineering Sciences, Vol 133, © 2021 WIT Press
www.witpress.com, ISSN 1743-3533 (on-line)
doi:10.2495/MC210081

2 MATERIALS AND METHODS

The fly ash (FA) used is filter residue produced by the treatment of fumes from the municipal solid waste incineration plant in Acerra (Naples). They are classified as hazardous waste in the European Waste Catalog, and cannot be used or even landfilled without prior treatment [4]. Therefore, a washing pre-treatment was carried out on FA in order to mainly reduce the content of heavy metals, chlorides and sulphates at the end of using the FA in the granulation process, which can be seriously compromised by the high content of soluble salts. For this purpose, the pre-treatment was carried out by washing in two phases with deionized water, each of 1.5 and led to a reduction of soluble salts, specifically a reduction of chlorides equal to about 67% and of sulphates 25%. After pre-treatment, the pre-treated fly ash was collected through a filtration process, and dried at $105 \pm 5°C$ in an oven for 24 h.

In the present work, the granulation technique was performed to stabilize FA, using various mix-designs of a ternary blend for the production of lightweight artificial aggregates. The production of the granules was carried out by means of a granulator on a pilot scale (Fig. 1), equipped with a rotating and tilting plate (d = 80 cm) for which the rotation speed and the angle of inclination were respectively set at 45 rpm for the rotation speed and 45 for the angle of inclination.

Figure 1: Granulator.

During the single granulation process, different mixtures are made, in which the FA is present at 80%, while the mass of limestone cement and granulated blast furnace slag vary respectively by 15%, 10% and 5%. In addition to the traditional single-pass granulation, a second granulation step was carried out. In the single-pass granulation the waste is incorporated in a binder matrix, in the double granulation on the other hand, the second phase is performed with 70% of cementitious binder, to obtain encapsulated aggregates inside an outer shell, to improve the technological properties and of leaching, in order to reach satisfactory levels of immobilization [5], [6]. Three different mixtures (A, B, C) were made for the production of single granulation granules (Granules (I)):

1. Granules A (I): FA 80%, CEM 5%, GGBFS 15%;
2. Granules B (I): FA 80%, CEM 10%, GGBFS 10%;
3. Granules C (I): FA 80%, CEM 15%, GGBFS 5%.

The double granulation process was performed starting from the aggregates obtained from the single granulation process with the aid of a new mixture composed of 30% limestone cement (CEM II/AL 42.5R) and 70% marble sludge (MS) in order to increase the amount of reused waste (marble sludge) and increase the thickness of the aggregates, previously made. In the double granulation process, the granules obtained from the single granulation are subjected to double granulation (Granules (II)):

1. Granules A (II): Granules A (I); CEM 30%; MS 70%;
2. Granules B (II): Granules B (I); CEM 30%; MS 70%;
3. Granules C (II): Granules C (I); CEM 30%; MS 70%.

The light artificial aggregates obtained from the granulation process (Fig. 2) were cured for 28 days at room temperature. This curing phase favors the hardening of the granules, which is necessary for subsequent manipulations and in such a way that it significantly improves their technological properties. All the systems listed underwent a characterization in terms of physico-chemical and mechanical properties.

Figure 2: Single granulation granules (Granules (I)).

3 RESULTS AND DISCUSSIONS

3.1 Chemical-physical-mineralogical characterization of the precursors

The granulation process uses fly ash (FA), granulated blast furnace slag (GGBFS) and limestone cement (CEM II/A-L 42.5R) as components of the systems, for the production of lightweight artificial aggregates. In addition, an innovative additional granulation phase was performed with cement binder and marble sludge (MS) in order to achieve satisfactory immobilization levels. The fly ash, granulated blast furnace slag, limestone cement and marble sludge before being used in the granulation process, were characterized from a chemical-physical-mineralogical point of view, using the following analytical techniques: X-ray fluorescence (XRF), X-ray diffraction (XRD), scanning electron microscopy (SEM) and particle size analysis.

The chemical composition of the FA as such and of each constituent of the mixtures, that is, treated FA, GGBFS, MS and cement was determined by X-ray fluorescence, and reported in Table 1.

Table 1: Chemical composition in terms of equivalent oxides (wt%) of fly ash as they are (FA TQ), treated fly ash (FA), granulated blast furnace slag (GGBFS), marble sludge (MS) and limestone cement (CEM II/A-L 42.5R).

Chemical composition (wt%)	FA TQ	FA	GGBFS	MS	CEM
Fe_2O_3	0.86	1.16	25.53	1.35	3.41
CaO	24.31	39.90	17.48	51.92	67.16
CO	16.35	14.13	11.29	22.74	–
SiO_2	2.62	4.57	16.26	14.16	16.65
Al_2O_3	1.53	3.36	8.93	4.56	4.21
SO_3	8.57	9.58	–	–	5.34
MgO	1.09	2.57	7.94	1.21	1.71
Mn_2O_3	–	–	3.44	–	–
Cr_2O_3	–	–	1.84	–	–
NO_x	–	–	10.07	–	–
SnO_2	–	–	–	2.20	–
Na_2O	13.87	9.10	–	0.86	–
K_2O	6.41	2.32	–	1.02	1.54
TiO_2	0.36	0.98	–	–	–
ClO	21.20	8.77	–	–	–
ZnO	2.85	3.36	–	–	–

The examination of the chemical analysis was a fundamental step for understanding the chemical matrix of the precursor, in fact, from the result of the X-ray fluorescence analysis (XRF), of the fly ash as such, it is possible to notice a presence of CaO higher than the other elements, in particular with respect to the SiO_2 and Al_2O_3 content. This abundance may be, in the first instance, due to the type of waste that was fed to the combustion chamber and, moreover, linked to the presence of Ca in the plant, since calcium-based materials are used in acid gas abatement systems. Also significant are the fraction of chlorine and sulphur equal to approximately 21% and 9% respectively. From the XRF analysis of the treated fly ash, it appears that the pre-treatments play a fundamental role by potentially reducing the soluble salts, and in particular the presence of chlorine which is approximately 9%.

Furthermore, it was possible, by means of X-ray diffraction analysis (XRD), to identify in the FA as they are, the presence of crystalline species containing calcium, such as, Lime (CaO), Calcite ($CaCO_3$), Portlandite (Ca $(OH)_2$), Calcium sulphate (Ca $(SO)_4$), and chlorine-containing phases such as sodium chloride and potassium chloride (NaCl and KCl), calcium chloride hydrate CaClOH. The identification of these crystalline species strengthens the XRF chemical analysis, in which a high percentage of calcium and chlorine is found compared to the other elements. Another crystalline species easily distinguishable from the diffraction spectrum is SiO_2. The treated fly ash has some crystalline phases found in fly ash as such as, Calcite ($CaCO_3$), Portlandite (Ca $(OH)_2$) and Calcium sulphate (Ca $(SO)_4$), in addition there are also other hydrated crystalline species containing calcium such as Syngenite (K_2Ca $(SO_4)_2 \cdot H_2O$), Gypsum ($CaSO_4 \cdot 2H_2O$), Calcium chloride dehydrate (Ca (ClO)$_2$·3H$_2$O). The presence of these last mineralogical phases is justifiable given the pre-treatment of the ashes, carried out by washing with water. Another difference that can be seen from the mineralogical analysis of fly ash as it is and those treated is that in the latter there are fewer crystalline phases containing chlorine (NaCl).

SEM micrographs were made for the samples of fly ash and treated fly ash, which showed the presence of a rough and irregular surface. Finally, a particle size analysis was carried out. Table 2 shows the particle size distribution of the fly ash, from which it can be seen that the particles of the treated FA are larger than the untreated ones.

Table 2: Particle size distribution of fly ash as they are (FA TQ), treated fly ash (FA).

Granules distribution (%)	FA TQ	FA
<1 µm	0	0
1–10.5 µm	5.84	4.75
10–20 µm	9.28	6.30
20–49 µm	29.35	16.73
49–80 µm	13.66	8.89
80–120 µm	8.29	6.03
>120 µm	33.58	57.30

The remaining components of the mixture, that is, GGBFS, MS and cement, in addition to being characterized by X-ray fluorescence (XRF), whose compositions in terms of equivalent oxides are reported in Table 1, were also characterized by diffraction analysis X-ray (XRD), scanning electron microscopy (SEM) and particle size analysis.

The results of the XRD analysis of limestone cement show that the main crystalline phase is Alite (Ca$_3$SiO$_5$) which gives the material a development of its resistance in the short term; followed by Belite (Ca$_2$SiO$_4$), which performs a function similar to halite but acting over the long term. Furthermore, these calcium silicates, reacting with water, form Portlandite, a crystalline phase with a Ca (OH)$_2$ structure, and an amorphous phase of hydrated calcium silicate, commonly known as C–S–H, responsible for setting and hardening of the paste. There are further crystalline phases such as Calcite (CaCO$_3$), Gypsum (CaSO$_4$ 2H$_2$O), Dolomite (CaMg (CO$_3$)$_2$) and Brownmillerite (phase containing iron and aluminium Ca$_2$ (FeAl) 2O$_5$). To ensure the principle of identity for the mineralogical components, even the blast furnace slag (GGBFS), waste produced by the steel industry, was analyzed using the X-ray diffraction (XRD) method. The mineralogical phases present in GGBFS are Wustite (FeO), Monticellite (Ca (FeMg) SiO$_4$), Gehlenite (Ca$_2$Al (AlSiO) OH). Finally, the crystalline phases present in the marble sludge (MS) are Quartz (SiO$_2$), Calcite (CaCO$_3$), ((KNa) (ASi$_3$O$_8$)) and Annite (KFeAlSi$_3$O$_{10}$ (OH)$_2$). Fig. 3 shows respectively the diffractogram and the mineralogical components detected in the cement, blast furnace slag (GGBFS) and marble sludge (MS).

The electronic scanning micrographs were also made on concrete, blast furnace slag and marble sludge. SEM micrographs of the cement at higher magnifications show the presence of particles with a rough and irregular surface and the presence of crystalline phases. The slag from the blast furnace has a rough and porous surface where it is possible to see a very extensive amorphous matrix and a minor presence of crystalline phases. For the marble sludge it is possible to note how the sample has a fine particle distribution, micrographs at higher magnifications show the presence here too of a rough and irregular surface, and the presence of crystalline phases, as shown in Fig. 4.

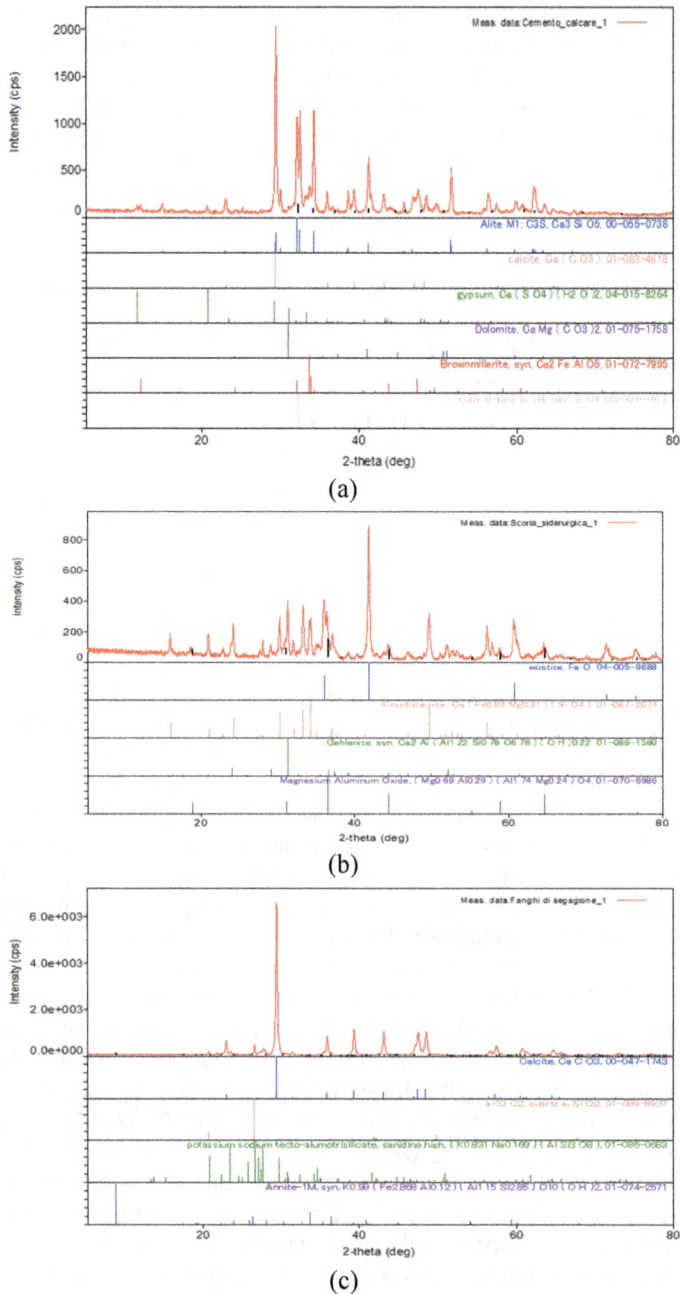

Figure 3: X-ray diffraction analysis (the reflection angle 2θ on the x axis and the intensity on the y axis) of (a) limestone cement (CEM II/A-L 42.5R); (b) granulated blast furnace slag (GGBFS); and (c) marble sludge (MS).

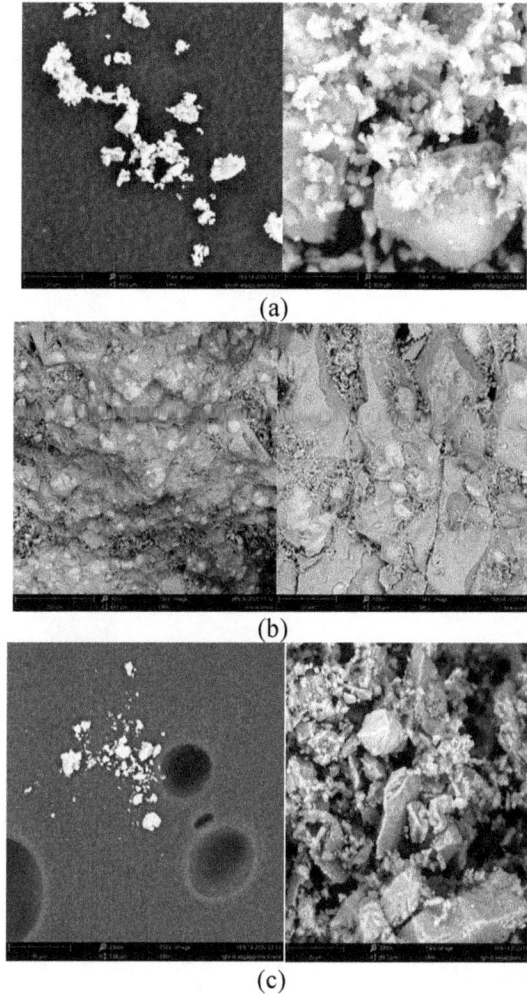

Figure 4:   Scanning electron microscopy of (a) limestone cement (CEM II/A-L 42.5R); (b) granulated blast furnace slag (GGBFS); and (c) marble sludge (MS).

Among the numerous factors influencing the granulation process, the initial particle size distribution is one of the most important properties of the raw material. In this regard, the absolute particle size distribution of blast furnace slag (GGBFS) and marble sludge (MS) is reported. GGBFS exhibits a monomodal absolute particle size distribution and the representative diameter of the particle population is 33.2 μm. The marble sludge appears to have a monomodal particle size distribution and the representative diameter of the particle population is 15.5 μm. In addition to the absolute particle size distribution, Fig. 5 also shows the cumulative particle size distribution of GGBFS and MS respectively. From the cumulative particle size distribution, it is possible to deduce another characteristic parameter, namely the median, respectively for GGBFS d50 = 33.15 μm and for MS d50 = 13.5 μm.

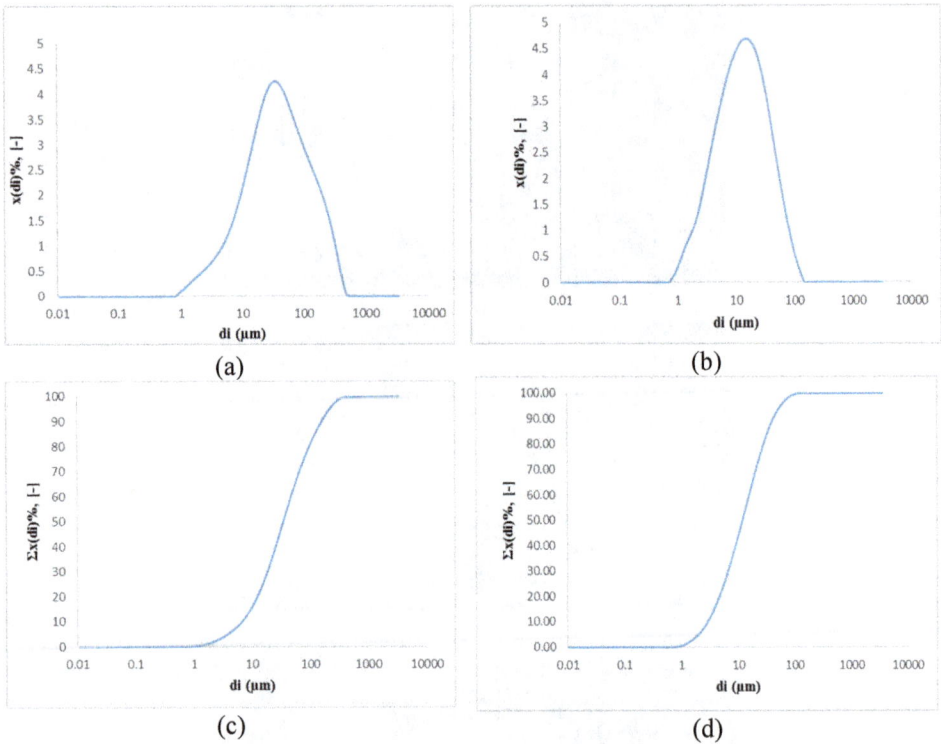

(a)

(b)

(c)

(d)

Figure 5:   Absolute particle size distribution (on x axis the values of the weight fraction of each particle size $x(di)$ and on y axis the values of the particle diameter di) of (a) granulated blast furnace slag (GGBFS); (b) marble sludge (MS) and cumulative particle size distribution (on x axis the sum of the values of the weight fraction of each particle size $x(di)$ and on y axis the values of the particle diameter di) of (c) granulated blast furnace slag (GGBFS); and (d) marble sludge (MS).

3.2 Chemical-physical and mechanical characterization of light artificial aggregates

3.2.1 Particle size analysis

After 28 days of curing, the light artificial aggregates obtained by a single and double granulation process were characterized in terms of physic-chemical and mechanical properties. Granules with dimensions ranging from 2 to 20 mm were obtained. The particle size distribution of the granules was determined according to the procedure described in the UNI EN 933-1 standard. In order to determine the particle size distribution of Granules A (I), Granules B (I), Granules C (I), Granules A (II), Granules B (II), Granules C (II) respectively, a sieving was carried out (or screening), a method used to establish the particle size distribution of a granular solid. The particle size analysis by sieving was performed by means of special sieves arranged in series, in the case in question of 2 mm, 4 mm, 8 mm, 10 mm, 16 mm, 20 mm, each of which retains the fraction of solid whose granules they are larger than the sieve holes. As regards the particle size distribution of Granules A (I), there is a higher concentration of granules passing between the sieves with an opening of 4 mm, 10 mm and 16 mm. Results similar to the previous case were also observed for Granules B (I). Finally, also for Granules C (I) it is possible to note that there is a higher concentration of granules passing between the sieves with an opening of 8 mm and 16 mm, including the sieve with an opening of 10 mm. Specifically, a greater presence of granules with dimensions greater than 10 mm of opening of the sieve was observed. Furthermore, only for Granules C (I), there are no granules having a diameter equal to 2 mm. This factor could be caused by the greater quantity of binder used, resulting in better agglomeration of the powders compared to the other two mixtures and the formation of granules with larger diameters. As regards the particle size distribution of granules A (II), in general, the percentage of the different diameters is well distributed. In particular, granules having dimensions greater than 16 mm of opening of the sieve, appear to be present in greater quantities, compared to the other granulometric classes. Furthermore, comparing these results with those obtained for Granules A (I), a substantial increase in the average size of the diameters is observed, in accordance with the single granulation process. Finally, from the results of the particle size distribution for Granules B (II) and Granules C (II), it is possible to make the same considerations made for Granules B (I) and Granules C (I), i.e. that there is a quantity greater than granules with a diameter of 16 mm and that substantially the rest of the particle size is well distributed.

3.2.2 Impact test

The impact test was performed on lightweight artificial aggregates, which is used to determine the impact factor of the aggregates and select them for a specific application. The impact test, in accordance with the UNI 12620-4 standard, consists in introducing the granules for a particle size class between 10–14 mm in a cylindrical container (d = 10 cm), which fills up to half. The apparatus consists of a metal sliding rod ended with a round point of 1.6 mm diameter, mounted in a suitable frame. The cylinder containing the granules is positioned inside the press, the piston is dropped 15 times, and a load of 8.9 kN is applied to the test sample. The test sample is collected and extracted from the cylinder. Using a 2 mm sieve, proceed to pass all the material below this dimension and proceed with the measurement of the weight determining the percentage of passage. Below are the results of the impact test carried out on the granules made in this report (Table 3); it should be remembered that this test allows to determine the percentage in heap of the granules which serves to define the final destination of the granular aggregates. In particular, according to the quantity of material passed through the 2 mm sieve, they are classified as: <15%, extremely strong; between 15% and 45%, satisfactory for road paving; >45%, extremely weak.

Table 3: Impact test results on single granulation granules (Granules (I)) and double granulation granules (Granules (II)).

	Percentage passing through the 2 mm sieve (%)
Granules A (I)	30.56
Granules B (I)	22.22
Granules C (I)	25.00
Granules A (II)	19.44
Granules B (II)	25.00
Granules C (II)	44.44

As can be seen from the results shown in Table 3, all the granules are satisfactory for use as a filling for road pavement, according to the UNI 12620-4 standard. In principle, the impact resistance is higher in Granules C (II), i.e. granules with 15% cement, 5% blast furnace slag and 80% FA, with the further addition of 30% cement and 70% sludge.

3.2.3 Compression test

With the compression test, the compressive strength which represents the maximum value of the applied stress is determined. In this study, compression tests were carried out on granular aggregates, in accordance with the UNI EN 12390-3 standard, using a Controls hydraulic press from Matest, on single and double granules with a diameter of 12.5 mm. Below are the results of the compression test on the granules, carried out in this report (Table 4).

Table 4: Press test results on single granulation granules (Granules (I)) and double granulation granules (Granules (II)).

	(MPa)
Granules A (I)	1.33
Granules B (I)	1.45
Granules C (I)	1.86
Granules A (II)	1.95
Granules B (II)	5.36
Granules C (II)	10.94

From the results of the granules, obtained from the single granulation process, it emerges that the breaking load, on the other hand, is almost constant, varying from a minimum of 1.33 MPa to a maximum of 1.86 MPa for the granules obtained from the single granulation process. The results of the granules obtained from the double granulation process are even more satisfactory than the aggregates obtained from the single granulation. In fact, by comparing all the values of the breaking loads, it is evident that the double granulation aggregates all have higher breaking loads than the corresponding aggregates obtained from the single granulation. This is because the granules obtained from the double granulation process have a greater quantity of binder, or 30% more by weight of cement, therefore, the second granulation phase determines an increase in compressive strength. As a result, the double granulation is more effective, in terms of mechanical performance, than the single granulation.

3.2.4 Assignment test

Release tests are tests in which a solid material is placed in contact with a liquid (leaching agent), resulting in a liquid product (eluate). The chemical analysis of the eluate allows to determine the chemical species that are released from the solid material over time, also contextualizing the potential danger to the environment. These tests were carried out for the granules, following the procedures described in the UNI 10802 standard. The samples were placed in contact with demineralized water, in liquid/solid ratio = 10, and left under stirring for 24 h. The solid residue was then separated by filtration, and the eluate obtained was analyzed on the atomic absorption spectrometer (AAS) for the determination of heavy metals. Furthermore, the leaching solution was analyzed by ionic liquid chromatography for the determination of chlorides and sulphates. Table 5 shows the results relating to the release of heavy metals, chlorides and sulphates (in mg L-1) of the granules deriving from the single and double granulation process and the relative limit values for non-hazardous waste.

Table 5: Results of the release test for single granulation granules (Granules (I)) and double granulation granules (Granules (II)).

	Cu	Pb	Zn	Cd	Cr	Chloride	Sulphates
Granules A (I)	0	1.15	1.11	0.01	0.45	39,439	7,283
Granules B (I)	0	0	0.28	0.04	0.1	22,930	5,158
Granules C (I)	0	0.71	0.32	0.06	0.55	22,996	14,859
Granules A (II)	0	0.49	0.62	0.03	0.49	9,819	4,601
Granules B (II)	0	0	0.17	0.01	0.35	7,101	909
Granules C (II)	0	0.24	0.29	0	0.53	13,333	3,595
Law limits[a]	5	1	5	0.1	1	1,500	2,000

[a]UNI 10802.

As for the results relating to the concentration of heavy metals, the only value that does not comply with the legislation is the concentration of Lead for Granules A (I). However, it should be noted that the value deviates slightly from the regulatory limit. Conversely, the leaching values of chlorides and sulphates were beyond the limits for almost all granules, except for Granules B (II) where the resulting sulphate concentration was below the limit. Also in this case, the double granulation process has shown benefits and it can certainly be said that the new conferment of thickness is able to better trap contaminants.

4 CONCLUSIONS

The irreversible and increasingly worrying crisis, together with environmental destruction, has led researchers to seek alternative solutions to natural resources. In particular, the aim is to promote the use of recycled aggregates to protect and support natural resources. Many studies have mainly confirmed the feasibility and efficiency of using recycled aggregates as building materials to reconsider the appropriateness of reusing industrial waste. In this regard, this study was conducted, from which it is possible to state that the light artificial aggregates produced by double granulation have brought numerous environmental and economic advantages, with the addition of a further waste (marble sludge) to give the new thickness. This translates into lower cost of waste disposal, lower cost of product preparation and less space in landfills. The results in terms of mechanical properties, deriving from this study, demonstrate the effectiveness of using various wastes in the production of aggregates through the double granulation process. At the same time, the results of the leaching tests showed a significant global release of the soluble salts. However, leaching from granules obtained by double granulation was reduced for both chlorides and sulphates.

REFERENCES
[1] Ren, P., Ling, T.C. & Mo, K.H., Recent advances in artificial aggregate production. *J. Cleaner Prod.*, 125215, 2020.
[2] Chen, J., Yang, D., Tang, W. & Wang, S.Y., Producing synthetic lightweight aggregates from reservoir sediments. *Constr. Build. Mater*, **28**, pp. 387–394, 2012.
[3] De Gisi, S., Chiarelli, A., Tagliente, L. & Notarnicola, M., Energy, environmental and operation aspects of a SRF-fired fluidized bed waste-to-energy plant. *Waste Manage*, **73**, pp. 271–286, 2018.

[4] Loginova, E., Proskurnin, M. & Brouwers, H.J.H., Municipal solid waste incineration (MSWI) fly ash composition analysis: A case study of combined chelatant-based washing treatment efficiency. *J. Environ. Manage*, **235**, pp. 480–488, 2019.

[5] Colangelo, F., Messina, F. & Cioffi, R., Recycling of MSWI fly ash by means of cementitious double step cold bonding pelletization: Technological assessment for the production of lightweight artificial aggregates. *J. Hazard. Mater*, **299**, pp. 181–191, 2015.

[6] Molino, B., De Vincenzo, A., Ferone, C., Messina, F., Colangelo, F. & Cioffi, R., Recycling of clay sediments for geopolymer binder production. A new perspective for reservoir management in the framework of Italian legislation: The occhito reservoir case study. *Materials*, 7(8), pp. 5603–5616, 2014.

USE OF POLYETHYLENE AS A FEEDSTOCK FOR VALUE ADDED PRODUCT RECOVERY: WAX RECOVERY FROM PYROLYSIS

SULTAN MAJED AL-SALEM

Environment & Life Sciences Research Centre, Kuwait Institute for Scientific Research, Kuwait

ABSTRACT

One of the main engineering objectives nowadays is to develop and design a sustainable practice whilst operating a technology that can yield value added products and be environmentally friendly in the same time. No better example can be provided of ongoing research of such technologies other than waste to fuel and energy solutions. Such technologies can recycle solid waste whilst maintaining a sustainable practice and operation. A prime example of such is the technique of pyrolysis, where waste is deteriorated in inert environments and products of such process are used in the place of petroleum fossil-based ones. One of the products of pyrolysis namely when using a polyolefin polymer is chemical waxes that were proven recently as a valuable product on the market. These waxes have a potential of having fuel range hydrocarbons and their market is of growing capacity as of late. In this work, the pyrolysis of polyethylene (PE) sourced as a virgin extrusion grade and used a translucent pellets, was used to produce waxes in the pyrolysis operation of fixed bed batch type. The pyrolysis reaction temperature took place between 500 and 800°C and waxes were collected and characterised for their fuel potential amongst their physical and rheological properties. The elemental carbon in the wax ranged between 81.04% to 87.67% showing high potential of carbon based material similar to the one of the original high density polyethylene (HDPE) feedstock material. Furthermore, the dynamic viscosity (at 40°C) ranged between 31 and 171 Centipoise. It is worth noting the that the fuel range of the wax was of high range of some 70% diesel fuel for material collected at 500°C which is attributed to the slow pyrolysis regime and the branched nature of the feedstock. This points towards utilising this product as a potential fuel source in addition to further upgrading instead of classical refinery ones.

Keywords: polyethylene, pyrolysis, diesel, energy, fuel.

1 INTRODUCTION

Thermal degradation in inert environments aimed at recovering value added products from plastics, has been classed at the top of the hierarchy of organic and plastic waste management technologies [1]. These technologies include pyrolysis whereby the material is treated in moderate to high temperatures (350–900°C) [2], in order to produce light non-condensable gases, oils, solid char and wax [3]. Various technologies and reactor set-ups are used to perform pyrolysis such as fluidised bed reactors (FBRs), conical spouted beds, fixed beds, screen reactors and rotary kilns [4]–[9]. Depending on the residence time and operating conditions, pyrolysis could also be performed in slow, flash or ultra-mode of operation [2], [10]. The primary product of the pyrolysis of polyolefin plastics is waxes formed as a result of the deterioration of the feedstock which for energy minimisation purposes to avoid further cracking, can be utilised as a standalone product specially in slow mode of operation [10]–[15]. Pyrolysis wax encompasses various advantages such as high embodied energy content and fuel range chemicals, ease of handling and storage; the use as a raw feedstock material in refining and petrochemical complexes for fuel production by cracking [13]–[15].

Due to the aforementioned reasons, attention to pyrolysis wax has been renewed nowadays as evidently so in R&D circles [14], [15]. This is also coupled with the fact that

WIT Transactions on Engineering Sciences, Vol 133, © 2021 WIT Press
www.witpress.com, ISSN 1743-3533 (on-line)
doi:10.2495/MC210091

EU legislations have also started to influence the world over in managing plastic waste and diverting it from landfill sites [16]. The majority of plastic waste is comprised of polyolefins (62%) which has been accumulating around the world over due to the latest surge in demand associated with the COVID-19 pandemic [15], [17]. In this work, the pyrolysis of polyolefin plastic grade in the form of high-density polyethylene (HDPE) took place to produce waxes (pyro-wax) in a novel and patented fixed bed batch reactor. The pyrolysis reaction temperature took place between 500 and 800°C and waxes were collected and characterised for their fuel potential amongst their physical and rheological properties.

2 EXPERIMENTAL

HDPE was acquired as a virgin film extrusion grade (Equate Co.). The polymer was used as received in the form of pellets with a measured size of approx. 3 ×3 mm. The reported density, melt flow index and melting point as declared by manufacturer are 0.952 g·cm^{-3}, 10 g/10 min and 131°C, respectively. The thermal profile of the feedstock is shown elsewhere [18]. The amount of 100 g of the virgin HDPE was subjected to thermal (non-catalytic) pyrolysis using a patented fixed bed reactor system with details shown previously in Al-Salem [6]. Wax yield was investigated between 500 and 800°C as an average of the three bed temperatures used in the operation. Slow pyrolysis was conducted and the wax collection was achieved post condensation (<3 h) without separating the heavy and light fractions using a heating rate of 5°C min^{-1} throughout.

The density of the wax produced was determined via specific gravity method as per the ISO 1183-1 namely using the immersion method-A. The wax sample was preheated to its pour point in a plastic cup and the density was calculated with respect to the specimen's apparent weight in air. Calorific Value (CV) was estimated following ISO 1928:20 by subjecting 0.5 g specimen to CAL3K Advanced bomb calorimeter under 3,000 kPa. The elemental analysis was also estimated for the wax samples. Duplicate 3 mg samples were subjected to a Perkin Elmer 2400 CHNS/O organic elemental analyser with an auto-sampler under reaction temperature varying between 925 and 1000°C depending on the element tested. The oxygen analysis of the instrument was calibrated using 1–3 mg of benzoic acid. Viscosity of waxes was determined using a cone and plate viscometer (HAAKE tester VTiQ) fitted with Peltier Temperature Controller. The dynamic viscosity value at 400 1 s^{-1} shear rate. Fuel range chemicals were identified using an Agilent 8,860 gas chromatography (GC) coupled with Agilent 5977B MSD mass spectrometer (MS) (HP5MS-UI Column, 30 m length, Stationary phase layer – 0.25 microns). Solid phase extraction was used for the processed analytes further on for characterisation.

3 RESULTS AND DISCUSSION

As mentioned in the first part of this communication, wax is the primary product evolved from the pyrolysis of polyolefin polymers namely polyethylene (PE). The devolitilisation that occurs in pyrolysis mechanism results in the residue that is called wax with a carbon range between C_{20} and C_{50} [17]. Wax is also responsible for adding to the gases and oils that accumulate as other pyrolysis products in the same mechanism [19]–[21]. Fig. 1 shows the yield of the wax formed and collected as an average of the duplicate experimental runs conducted in this work. With the exception of the 600°C experimental run, the yield was decreasing as a function of the reactor's temperature.

HDPE is a branched semi-crystalline polymer that is characterised as a lesser branched material than low density polyethylene (LDPE) and more so when compared to linear low-

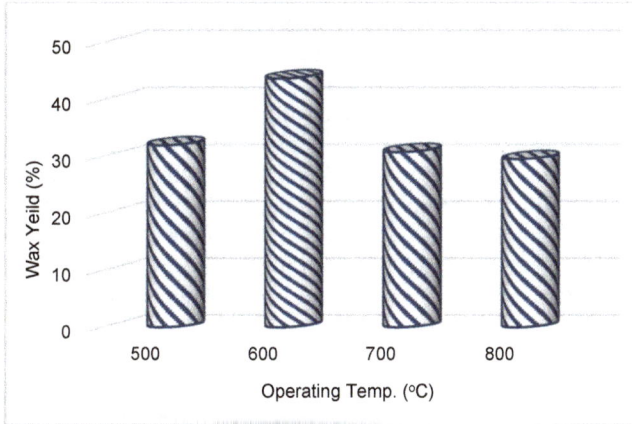

Figure 1: HDPE yields (wt.%) of wax with respect to reactor operating temperature.

density polyethylene (LLDPE). The cracking of wax was less evident when more heat (in the form of elevated temperature) at 500°C. This shows an estimated wax yield of some 32% (Fig. 1). Moderate synergy between the evolved gases and waxes was observed at 600°C to yield the highest wax (43%). Previous work conducted by other authors on PE using various set-ups show that wax yield reaches some 50% at 500°C with a gradual decrease [10], [22]. Fig. 2 shows the density of the wax obtained as a function of the reactor's temperature. A gradual decrease was observed in the analysed samples with an average density equal to 849.4 ± 5.32 kg·m^{-3}. The obtained density range is an agreement with previous reports on waxes and were higher than commercial wax (paraffin wax on the market) [23].

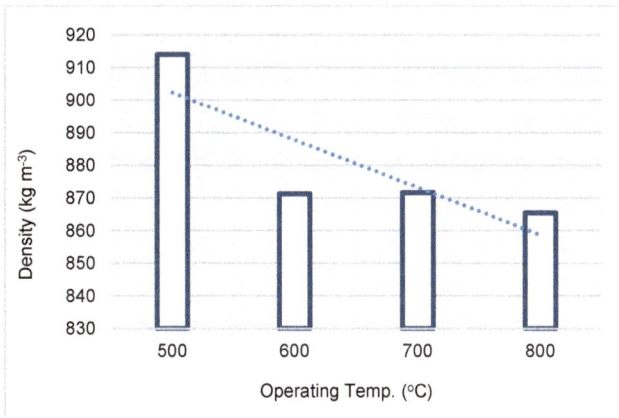

Figure 2: Density of wax (kg m^{-3}) with respect to reactor operating temperature.

The elemental analysis for the obtained wax as an average for all experimental runs was determined thus: Carbon (C, 79.46%), hydrogen (H, 17.77%) and sulphur (S, 2.77%). The calorific value (CV) was also estimated (average 46.22 kJ·g^{-1}) and is shown in Fig. 3.

There was no trend observed in the estimated values herein. It is quite important to highlight to aspects from the extracted work thus far. Firstly, commercial fuels have a CV range between 44.6 and 46.6 8 kJ·g^{-1} [24]. This shows that the wax in this work falls within said range promoting it as a potential fuel. Secondly, the wax has more than twenty-fold the S content that is permissible for diesel fuel engines fuels which also points towards the need for de-sulphurisation of the feedstock (wax) to obtain ultra-quality fuels using cracking. Fig. 4 shows the dynamic viscosity determined for the samples studied as a function of the reactor temperature. An incremental increase in a proportional manner was observed with the temperature. The range of the viscosity was between 31 and 171 centipoise falling in the range of waxy crude oil/asphaltene mix typically handled in refineries [25].

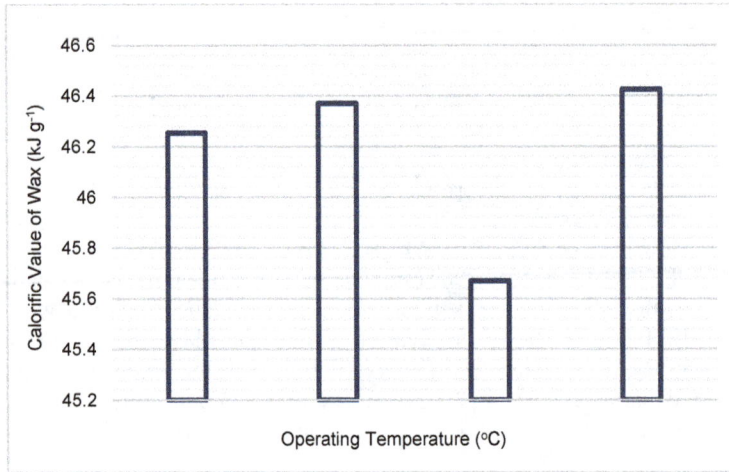

Figure 3: Calorific value of wax (kJ kg^{-1}) with respect to reactor operating temperature.

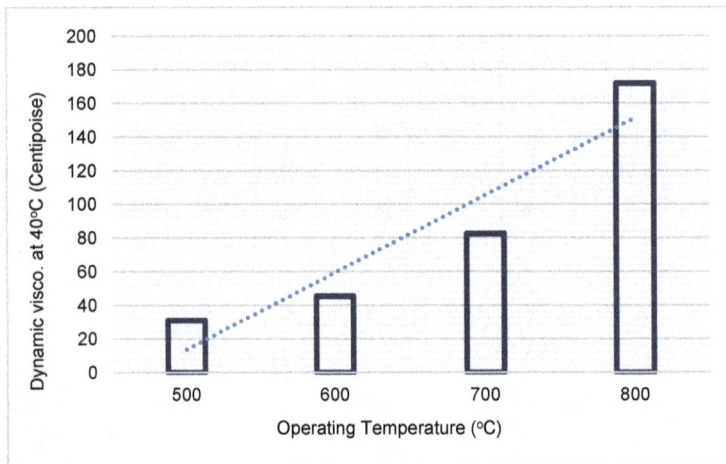

Figure 4: Dynamic viscosity of wax (centipoise) with respect to temperature.

Fig. 5 shows the chromatography of the studied samples by depicting the fuel range chemicals after characterising each carbon range as per the following: C_4–C_9 (petrol), C_{10}–C_{19} (diesel) and heavy waxes (C_{19}+). It is quite important to recognise the fact that the diesel fraction was increasing from about 15% at 500°C reaching 42.6% at 800°C. The concoction of the tested wax samples paves the way for future consideration of polyolefin waste that could substitute refinery feedstock for catalytic cracking. Fuel production could easily be achieved coupled with sulphur regulation that could reduce environmental impacts of plastic waste and dependence on crude oil sources of fuel.

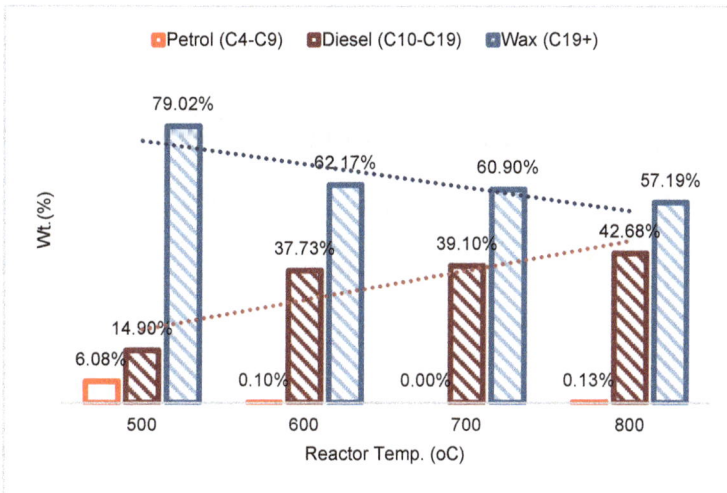

Figure 5:   Chromatographic analysis showing fuel range chemicals detected for wax recovered from the pyrolysis of HDPE with respect to temperature.

4 CONCLUSIONS

In this work, high density polyethylene (HDPE) was subjected to thermal pyrolysis in the range from 500 and 800°C. The wax yield was at a maximum at 600°C reaching some 43% of the total reactor charge. The experimental characterisation of the samples showed that the wax conformed with fuels available on the market. Furthermore, the sulphur content exceeds the permissible limit of 0.1 wt.% which points towards the fact that de-sulphurisation is required to reduce the S element content. In addition, the diesel fraction in the samples reached some 42.6% for samples extracted at 800°C. In general, the fuel range chemicals show possibility of fuel production from the wax samples by cracking which makes it a suitable feedstock for refineries and petrochemical complexes. The work presented also provides a platform for circularity in the future where integration opportunities are possible with the oil and gas sector at one end; and environmental and waste management services at the other.

ACKNOWLEDGEMENTS

The author would like to thank the Kuwait Foundation for the Advancement of Sciences (KFAS) for funding and supporting this research project through the Grant for Project EM096C (PN18-14SC-04). Gratitude is also expressed to the Kuwait Institute for Scientific Research (KISR) for internal fund acquirement and support.

REFERENCES

[1] Al-Salem, S.M., Lettieri, P. & Baeyens, J., Recycling and recovery routes of plastic solid waste (PSW): A review. *Waste Management*, **29**(10), pp. 2625–2643, 2009.

[2] Al-Salem, S.M., Antelava, A., Constantinou, A., Manos, G. & Dutta, A., A review on thermal and catalytic pyrolysis of plastic solid waste (PSW). *Journal of Environmental Management*, **197**, pp. 177–198, 2017.

[3] Sharuddin, S.D.A., Abnisa, F., Daud, W.M.A.W. & Aroua, M.K., A review on pyrolysis of plastic wastes. *Energy Conversation & Management*, **115**, pp. 308–326, 2016.

[4] Hita, I., Arabiourrutia, M., Olazar, M., Bilbao, J., Arandes, J.M. & Sánchez, P.C., Opportunities and barriers for producing high quality fuels from the pyrolysis of scrap tires. *Renewable & Sustainable Energy Reviews*, **56**, pp. 745–759, 2016.

[5] Al-Salem, S.M., Valorisation of end of life tyres (ELTs) in a newly developed pyrolysis fixed bed batch process. *Process Safety & Environmental Protection*, **138**, pp. 167–175, 2002.

[6] Al-Salem, S.M., Thermal pyrolysis of high density polyethylene (HDPE) in a novel fixed bed reactor system for the production of high value gasoline range hydrocarbons (HC). *Process Safety & Environmental Protection*, **127**, pp. 171–179, 2019.

[7] Sharma, B.K., Moser, B.R., Vermillion, K.E., Doll, K.M. & Rajagopalan, N., Production, characterization and fuel properties of alternative diesel fuel from pyrolysis of waste plastic grocery bags. *Fuel Processing Technology*, **122**, pp. 79–90, 2014.

[8] Kunwar, B., Cheng, H.N., Chandrashekaran, S.R. & Sharma, B.K., Plastics to fuel: A review. *Renewable & Sustainable Energy Reviews*, **54**, pp. 421–428, 2016.

[9] Jiang, G., Wang, J., Al-Salem, S.M. & Leeke, G.A., Molten solar salt pyrolysis of mixed plastic waste: Process simulation and techno-economic evaluation. *Energy & Fuels*, **34**, pp. 7397–7409, 2020.

[10] Aguado, R., Olazar, M., San José, M.J., Gaisan, B. & Bilbao, J., Wax formation in the pyrolysis of polyolefins in a conical spouted bed reactor. *Energy & Fuels*, **16**, pp. 1429–1437, 2002.

[11] Wax Market-Growth, Trends, and Forecast (2020–2025), 2019. http://mordorintelligence.com/industry-reports/waxes-market. Accessed on: 3 Apr. 2020.

[12] Yousef, S., Eimontas, J., Zakarauskas, K. & Striugas, N., Microcrystalline paraffin wax, biogas, carbon particles and aluminium recovery from metallised food packaging plastics using pyrolysis, mechanical and chemical treatments. *Journal of Cleaner Production*, **290**, p. 125878, 2021.

[13] Rodríguez, E., Palos, R., Gutiérrez, A., Trueba, D., Arandes, J.M. & Bilbao, J., Towards waste refinery: Co-feeding HDPE pyrolysis waxes with VGO into the catalytic cracking unit. *Energy Conservation & Management*, **207**, p. 112554, 2020.

[14] Qureshi, M.S. et al., Pyrolysis of plastic waste: Opportunities and challenges. *Journal of Analytical & Applied Pyrolysis*, **152**, p. 104804, 2020.

[15] Al-Salem, S.M., Yang, Y., Wang, J. & Leeke, G.A., Pyro-oil and wax recovery from reclaimed plastic waste in a continuous auger pyrolysis reactor. *Energies*, **13**, p. 2040, 2020.

[16] Renewable Energy Directive (2009/28/EC), European Commission, sustainability criteria, 2018. https://ec.europa.eu/energy/topics/renewable energy/biofuels/sustainability-criteria_en. Accessed on: 10 Mar. 2021.

[17] Arabiourrutia, M., Elordi, G., Lopez, G., Borsella, E., Bilbao, J. & Olazar, M., Characterization of the waxes obtained by the pyrolysis of polyolefin plastics in a conical spouted bed reactor. *Journal of Analytical & Applied Pyrolysis*, **94**, pp. 230–237, 2021.

[18] Al-Salem, S.M. et al., Non-isothermal degradation kinetics of virgin linear low density polyethylene (LLDPE) and biodegradable polymer blends. *Journal of Polymer Research*, **25**(5), p. 111, 2018.

[19] Conesa, J.A., Font, R., Marcilla, A. & Caballero, J.A., Kinetic model for the continuous pyrolysis of two types of polyethylene in a fluidized bed reactor. *Journal of Analytical & Applied Pyrolysis*, **40–41**, pp. 419–431, 1997.

[20] Conesa, J.A. & Font, R., Kinetic severity function as a test for kinetic analysis: Application to polyethylene pyrolysis. *Energy & Fuels*, **13**, pp. 678–685, 1999.

[21] Al-Salem, S.M. & Lettieri, P., Kinetic study of high density polyethylene (HDPE) pyrolysis. *Chemical Engineering Research & Design*, **88**(12), pp. 1599–1606, 2010.

[22] Hájeková, E. & Bajus, M., Recycling of low-density polyethylene and polyprylene via copyrolysis of polyalkene oil/waxes with naphtha: Product distribution and coke formation. *Journal of Analytical & Applied Pyrolysis*, **74**, pp. 270–281, 2005.

[23] Ukrainczyk, N., Kurajica, S. & Šipušiæ, J., Thermophysical comparison of five commercial paraffin waxes as latent heat storage materials. *Chemical & Biochemical Engineering Quarterly*, **24**(2), pp. 129–137, 2010.

[24] Williams, P.T., Pyrolysis of waste tyres: A review. *Waste Management*, **33**, pp. 1714–1728, 2013.

[25] Kriz, P. & Andersen, S.I., Effect of asphaltenes on crude oil wax crystallization. *Energy & Fuels*, **19**, pp. 948–953, 2005.

THERMAL AND FIRE BEHAVIOUR OF CEMENT BLOCKS WITH RECYCLED ROOF WASTES

RAQUEL ARROYO, LOURDES ALAMEDA, ALVARO ALONSO, SARA GONZÁLEZ,
VERÓNICA CALDERÓN, SARA GUTIÉRREZ & ÁNGEL RODRÍGUEZ
Universidad de Burgos, Spain

ABSTRACT
In accordance with the European politics of reducing the amount of plastics and polymers sited in landfills, the inclusion of compounds such as roof wastes as recycled and reusable materials to replace variable amounts of aggregates is interesting in the production of new construction materials due to their physical and chemical behaviour. Mortars made with Portland cement, sand, water and grinded roof wastes from the automobile industry that replace in different amounts part or all of the aggregates are examined in this study. To try to avoid the mechanical resistance limitation due to the use of roof wastes, the chemical properties of the binders have been modified with non-ionic surfactants that changed the effect on the hydration of the clinker. This variation produces an important change in the mechanical resistance to achieve recycled structural materials with a low density compared to conventional light mortars. In addition, these additives improve other properties including workability, compaction of the matrix, prevent the disintegration of the particles and help to improve the mechanical properties, ductility, thermal resistance and durability against fire to reinforce the materials. These eco-mortars have a lower thermal conductivity as more quantity of roof wastes are incorporated, which greatly favours the thermal insulation of the final envelope, as well as a good behaviour against temperature, measured in terms of thermogravimetry and non-combustibility test. With these results, we can consider the use of roof wastes as a sustainable alternative to the materials currently used and then with them we can be able to contribute to a more ecological business model in the building sector.
Keywords: recycled ceilings, lightweight prefabricated, fire resistance, polyurethane.

1 INTRODUCTION
In an effort to reduce the dependence on raw materials and reuse waste, we are working on numerous awareness, reuse and recycling actions in European, national and regional programs, which leads to the development of new techniques for recycling wastes for the construction materials, moving from waste to new raw materials, in order to close production cycles towards continuous reuse.

The polyurethane sector involves only in Europe 18,000 people and moves a turnover of about 4,000 million euros. Worldwide, it involves 240,000 companies, with one million jobs and generates an economy worth of about 207 billion euros. Trials are underway to introduce recovery systems for construction waste in order to divert it from landfills and treat it according to the other options at the end of its life. The main technologies for recycling polyurethane and its derivatives are energy recovery, mechanical recycling and chemical recycling. The lack of a collection, sorting and processing infrastructure has somewhat blocked the recycling of this waste.

In addition, the construction sector plays an important role in the economy. It generates almost 10% of GDP (Gross Domestic Product) and provides 20 million jobs in Europe, mainly in micro and small companies. Moreover, the materials used in building and civil works represent 42% of our final energy consumption, approximately 35% of our greenhouse gas emissions and more than 50% of all the materials removed.

This research is about the use of waste from complete roofs generated in the automobile industry, valued as raw material with added value for its incorporation into the precast

WIT Transactions on Engineering Sciences, Vol 133, © 2021 WIT Press
www.witpress.com, ISSN 1743-3533 (on-line)
doi:10.2495/MC210101

industry of the construction sector, less technical and consumer of a huge amount of natural resources (aggregates, cement, energy, water, CO_2 emissions, etc.)

The replacement percentage depends on the properties of the final product required in each case, with final properties in the hardened state enough to be able to apply the current legislation.

It is proposed as the basis for a circular economy network and reuse of a waste found in large quantities, and the development of sustainable innovation solutions for cement mortars in construction.

2 RAW MATERIALS

Prefabricated fire-resistant cement lightened with industrial waste of polymeric origin, from recycled vehicle roofs. We have manufactured it with the form of blocks.

The characteristics and nature of construction materials, based on cement, make possible to enhance foams that come from polyurethane insulating panels with remains of other materials (adhesives, metal oxides, remains of paint and/or plastering, etc.) to avoid the difficulties to reuse the material and prevent its shipment to landfill.

The precast is composed of the raw materials detailed below:

- Commercial cement;
- Arid;
- Shredded waste with polyurethane matrix;
- Additives;
- Water.

The cement/aggregate dosage is 1/6 by weight, considering the aggregate as the addition of sand plus the polymer residue. The precast cement lightened with polymer matrix residues, provides improvements in the physical characteristics of the precast, with weight reductions of up to 70%.

2.1 Cement

The cement is the CEM I 52.5 R. type according to the UNE-EN 197-1: 2011 "Cement" standard, Portland Cement with a mass composition of 95–100% of clinker and 0–5% of minority components. These values refer to the cement core excluding calcium sulphate and any additives. The mechanical requirements are right with a compressive strength at 2 days more than 30 MPa and at 28 days more than 52.5 MPa. The beginning of setting is more than 45 min and the expansion is less than 10 mm, which meets the physical requirements according to the regulations. The chemical requirements are also adequate, with loss on ignition less than 5%, insoluble residue less than 5%, sulphate content less than 4% and chloride percentage less than 0.10%.

2.2 Aggregates

The aggregates used in the preparation of mortars follow the standard UNE-EN 13139: 2003 "Aggregates for mortars", with the use of particles smaller than 4 mm and with a sand with rounded and not very angular shapes, typical of granular natural aggregates.

The content in fines for this 0/4 aggregate has a maximum percentage in passes through the sieve of 0.063 mm of 5%. The bulk density is 1,670 kg/m^3.

2.3 Polyurethane

The polyurethane waste comes from recycled vehicle roofs, with a bulk density of 92.5 kg/m^3 and a density of 1,681 kg/m^3.

Two different types of additives have been used (non-ionic surfactants, hydrophobic/hydrophilic composition) in the manufacture of mortars, which modifies the chemical properties of their components. It produces an improvement in the hydration of the cement (they reduce the amount of water needed in dosages) and improve their properties. We used a percentage of additives of 1% in relation to the weight of cement.

This alteration in the components of the precast cement helps to maintain their mechanical strength, obtaining recycled materials with structural properties, and at the same time, with a low density in relation to conventional light mortars.

2.4 Water

We added water in an amount that guarantee an appropriate consistency, good workability and a plastic state in the mixtures, in accordance with the UNE-EN 1015-3: 2000 standard "Test methods for masonry mortars. Part 3".

3 RESULTS

3.1 Properties in fresh and hardened state

We have mixed cement type CEM I 52.5 R, the crushed waste from recycled vehicle roofs with a polyurethane matrix (with particle sizes less than 4 mm) and aggregates (the amount of sand is replaced by different percentages of the polymer waste by volume). The cement/aggregate dosage is 1/6 by weight, considering the aggregate as the addition of sand plus the polymer waste. We added different additives to each dosage: one very hydrophilic and the other slightly hydrophilic to study their influence.

We made different dosages and measured the properties in fresh and hardened state.

1. Substitution of 50% of aggregate with polymer waste, and a hydrophilic unit 3 EO (very hydrophilic) in relation to the weight of cement.
2. Substitution 100% aggregate with polymer waste, and a hydrophilic unit 3 EO (very hydrophilic) in relation to the weight of cement.
3. Substitution of 50% aggregate with polymer waste, and a hydrophilic unit 10 EO (little hydrophilic) in relation to the weight of cement.
4. Substitution of 100% aggregate with polymer waste, and a hydrophilic unit 10 EO (little hydrophilic) in relation to the weight of cement.

The tests were carried out in accordance with the UNE-EN 1015 standard "Test methods for mortars for masonry", with all the requirements in all cases. For a good interaction, we mixed on the one hand, the cement, the water and the additive, to improve the effect of the surfactant on the cement.

Then, we added the mixture of waste and aggregate, and then we continue with the manufacture of the mixture with the conventional method.

The properties in fresh and hardened state are described in Tables 1 and 2.

Table 1: Properties in fresh and hardened state.

Dosages	Density (kg/m^3)	Shore hardness (Shore C)	Flexural strength (MPa)	Compressive strength (MPa)
1	1,667	81.3	5.38	13.27
2	873	46.7	2.52	4.05
3	1,250	84.5	2.85	5.13
4	763	57.1	1.62	3.71

Table 2: Properties in fresh and hardened state.

Dosages	Water absorption due to capillarity (kg/m$^2\cdot$min$^{0.5}$)	Mortar classification	Absorption
1	4.24	W0	12.2
2	9.95	W0	48.6
3	7.15	W0	24.2
4	11.35	W0	59.5

3.2 Non-combustibility test

With the mixes described above, large blocks were manufactured to be able to place on site in a simple way as prefabricated products. We made the dosages described in Table 3.

Table 3: Dosages for large blocks.

Dosages	Cement (kg)	Aggregate (kg)	Polymer waste (kg)	Water (kg)	Additive (g)
1	6.67	20	1.12	5.63	66
2	5	–	1.62	5.10	50
3	3.75	12.3	0.63	2.73	38
4	3.75	–	1.25	3.60	38

The above mixtures have been tested to the "Non-combustibility test" that is defined in the UNE EN-ISO 1182: 2011 "Reaction to fire tests of products". The results obtained are described in Table 4.

Table 4: Results obtained in the non-combustibility test.

Dosages	Oven temperature increase (Δt) (°C)	Persistence of the inflammation (t_f) (s)	Loss of mass (Δm) (%)
1	41.6	368	10.72
2	48.5	1,072	32.31
3	29.2	517	14.24
4	81.3	745	38.17

Taking into account only their contribution to the flammability of the materials, these results indicate that all mixtures except number 4 can be classified on fire as Euroclass A2, that is, non-combustible, without contribution to fire, according to the UNE-EN standard. 13501-1: 2007 + A1: 2010 "Classification based on fire behaviour of construction products and building elements. Part 1: Classification from data obtained in Reaction to fire tests".

4 CONCLUSIONS

We have made a precast concrete lightened with the shape of a block with industrial waste of polymeric origin from recycled vehicle roofs that is fire resistant. For this, residues have replaced different percentages of aggregate from the precast mortar. Different additives, with polymeric constitution, have also been added to study their influence. In this way, mechanical resistance is maintained and even improved in relation to the reference values required by European regulations.

The density decreases as residue is added in relation to reference precast currently on the market, which means savings in the base structure and better workability when placing them on site.

In addition, these eco-precast mortar have a lower thermal conductivity as residue is incorporated, which greatly favours the thermal insulation of the final product, as well as a good behaviour against temperature and against fire, measured in terms thermogravimetry and non-combustibility (reaction to fire).

ACKNOWLEDGEMENTS

Authors gratefully acknowledge the financial support of BU070P20 Project funded by the Fondo Europeo de Desarrollo Regional (FEDER) of the EU and the Junta de Castilla y León (Spain).

REFERENCES

[1] Arroyo, R., Horgnies, M., Junco, C., Rodríguez, A. & Calderón, V., Lightweight structural eco-mortars made with polyurethane wastes and non-ionic surfactants. *Construction Building Materials*, **197**, pp. 157–163, 2019.

[2] de Souza Kazmierczak, C., Dutra Schneider, S., Aguilera, O., Carine Albert, C. & Mancio, M., Rendering mortars with crumb rubber: Mechanical strength, thermal and fire properties and durability behavior. *Construction Building Materials*, **253**, p. 30, 2020.

PYROLYSIS OF END OF LIFE TYRES RECLAIMED FROM LORRY TRUCKS: PART I – OIL RECOVERY AND CHARACTERISATION

SULTAN MAJED AL-SALEM

Environment & Life Sciences Research Centre, Kuwait Institute for Scientific Research, Kuwait

ABSTRACT

About 1.5 billion tyre units are classed as a solid waste on an annual basis after their disposal. The carcass of end of life tyres encompass various petrochemical substances that could easily be recovered after thermally cracking the rubber fraction. In this work, experimental studies were carried out on end of life tyres (ELTs) reclaimed from lorry trucks which are known to be quite resilient to environmental exposure. The cracking took place between 500 and 800°C in a pyrolysis operation of fixed bed reactor type. Oil recovered was extensively studies for its properties and fuel potential. Sulphur content was also determined reaching 2.92% for oil extracted at 500°C with elemental carbon estimated at 64.78%. Elevated temperatures of operation have also shown high potential of diesel fuel fraction (C_{10}-C_{19}) in the pyrolysis oil where it reached 80% for oil recovered at 800°C.
Keywords: tyres, pyrolysis, diesel, energy, waste.

1 INTRODUCTION

It is estimated that the European Union (EU) alone produces some 289 million tyre units on an annual basis and that end of life tyres (ELTs) are estimated to be in the range between 1.3 and 1.5 billion units each year [1]. Some 64% of this total number of tyres discarded as a special category waste or rubber solid waste is still diverted to landfill sites [2]. According to the United States Environmental Protection Agency (EPA), 9.16 million tonnes of rubber waste was generated in 2018 and only 1.67 million tonnes were recycled in the same year [3]. Furthermore, ELTs are known to resist degradation due to exposure and environmental factors. Therefore, landfilled ELTs and stockpiled ones present an immediate danger to the environment, as well as residents to the areas close by such sites. ELTs disposed of in improper manner can lead to fire ball hazards, rodent and insect infestation; and environmental pollution [4].

The carcass of a tyre consists of natural rubber (10–30 wt.%), styrene-butadiene rubber (30–50 wt.%), butadiene rubber (30 wt.%) and carbon black (30 wt.%). Other additives (inorganics and organic ones) and sulphur make up some 1% of its constituent [2]. This supports the fact that ELTs embody long chain hydrocarbons similar to the ones present in fuels produced from refineries and petrochemical complexes. According to Hita et al. [5], lorry (truck) tyres consist of about 30 to 50 wt.% of natural rubber depending on either US or European standard which is higher than passenger car tyres.

There exist three main routes for ELTs management nowadays that can provide numerous advantages both on operational and environmental fronts. The first route of ELTs treatment is material recovery whereby the tyre is mechanically grinded using various techniques to reduce its size to 1 cm (or below). The product could be used in civil engineering applications, roads and coastal purposes as well [6]. The second means of ELTs treatment is via combustion with the aim of energy recovery [7]–[9]. Lastly, ELTs could be valorised using the technique of pyrolysis whereby the tyres are treated under inert atmospheres for fuel and light gas recovery to recover valuable products competing with one produced from fossil fuels [10]–[13]. Potential products and their characteristics could be found in detail elsewhere [1].

The problem of ELTs stockpiling and their management extends the world over. Moreover, the Middle-East and namely in Kuwait, the largest ELTs grave yard is present with over 7 million units stockpiled in one location [14]. In this work, lorry (truck) ELT grade was studied using a fixed bed reactor to achieve pyrolysis conditions with the aim of studying the pyrolysis oil (pyro-oil) quality properties and its potential as a fuel substitute. This part of the communication focuses on the oil extracted quality. The work presented herein can also pave the way for strategies in Middle-East and by extension other regions to have a circular economy basis using ELTs as a potential feedstock material with value added products recovered.

2 EXPERIMENTAL

ELTs carcasses (10 kg) were acquired from Al-Essa company (Kuwait) which were reclaimed ensuring similar make and model of each tyre grade. The ELTs were of lorry truck type reclaimed originally from Al-Maillam group dealership (315/80R22.5). Each tyre was firstly air blown to remove any dust particles and then subjected to shredding using ELDAN (Denmark) at 60°C. The obtained samples (manually measured) were 1.22 cm in size. The samples were stored in laboratory conditions (≈22°C) using sealed plastic containers. Feedstock in the amount of 200 g was placed in the fixed bed as a charge with Alumina packing of a 5 mm diameter (average bulk density of 700 kg m^{-3}) was used in the amount of 120 g. For operation and reactor specifications, the readers are referred to Al-Salem [12]. Oil yield was studied between 500 and 800°C as an average of the three bed temperatures used in the operation. Pyrolysis was conducted and the oil collection was achieved post condensation (<3 hours of pyrolysis time) using a heating rate of 5°C min$^-$ throughout. A CAL3K Advanced bomb calorimeter was used for gross calorific value (CV) determination of the oil samples as per ISO 1928:20 using 0.5 g specimens. Samples were analysed using a Perkin Elmer CHNS/O elemental analyser 2,400 organic elemental analyser with a MAS auto-sampler. Flash point was determined using a Koehler Closed Cup Rapid Flash Tester. The GCMS analysis was carried out using an Agilent 8,860 GC and an Agilent 5977B MSD. The samples went through a Solid Phase Extraction (SPE) procedure before being injected in the GC, in order to extract the chemicals of interest and avoid column contamination and saturation with heaviest chemicals. The fuel range chemicals were determined and categorised as per the following: Petrol (C_4-C_9), diesel (C_{10}-C_{19}) and wax (C_{19}+).

3 RESULTS AND DISCUSSION

Fig. 1 shows the yield obtained as an average of the experimental runs conducted. The oil was produced at a maximum (84 g) when 500°C operating temperature was used and decreased gradually reaching 68 g at 800°C. This could be attributed to the low temperature providing more residence time whereby the carbon content in the material is cracked more towards primary hydrocarbon oil formation. No waxes were produced throughout the experimental runs. This indicates that primary reaction was complete for all polymeric content in the ELTs carcass for the intermediate stages of products [15]–[17].

Fig. 2 shows the estimated CV as a function of the operating temperature in this work. It is quite clear that the CV was increasing incrementally as a function of the reactor's bed temperature from 41.2 to 42.1 kJ g^{-1}. The CV as an average falls within the range of previous reports by other authors using various reactor set-ups [1], [18], [19]. It is quite

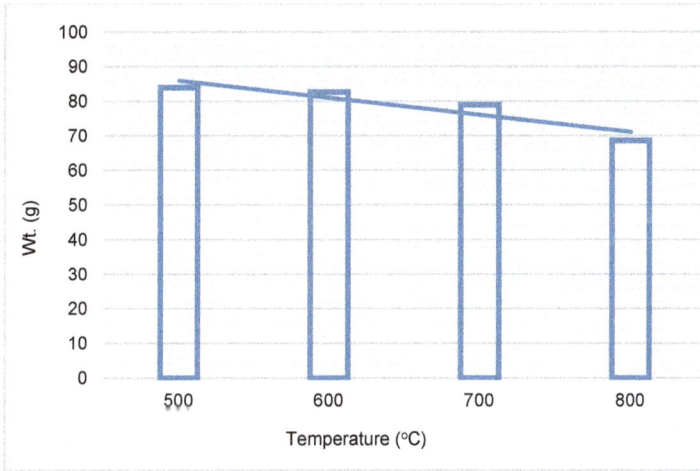

Figure 1: Oil yield (g) as a function of reactor temperature (°C).

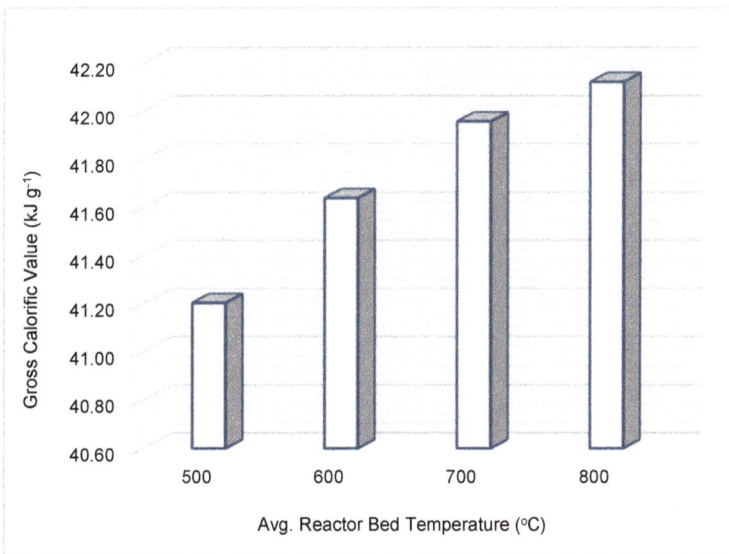

Figure 2: Calorific value (kJ kg^{-1}) as a function of reactor temperature (°C).

essential at this stage to point out the fact that the pyrolysis oil obtained herein are comparable to diesel fuel with almost similar CV [5]. The elemental analysis of the pyrolysis oil shows that the elemental carbon was at maximum for the 600°C (80.83%) whilst the sulphur was maximum for the 500°C (2.44%). De-sulphurisation is essential in this case since the sulphur content is quite high and requires to be reduced to 0.1% [20].

The flash point ranged between 40.1 and 42°C. The flash point indicates the flammability of the liquid as the lowest temperature which the substance is ignited with air [21]. Generally, pyrolysis oil has a lower flash point when compared to gas oil, kerosene or diesel [1]. Past reports on pyrolysis oil extracted from ELTs has ranged between 17 and 65°C [21]–[24]. The flash point of the pyrolysis oil extracted from truck tyres in this work has crossed the mark of kerosene. Fig. 3 shows the diesel fuel range estimated in the oil samples as a function of the operating temperature.

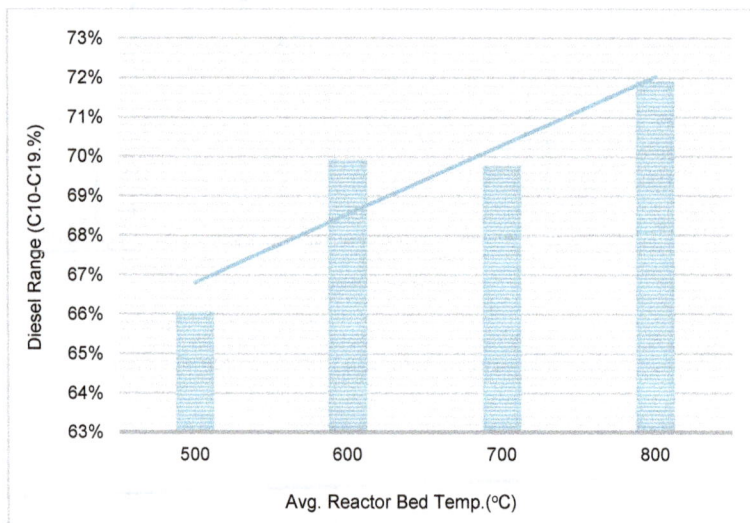

Figure 3: Diesel fraction (%) as a function of reactor temperature (°C) in oil samples studied in this work.

The diesel fuel was increasing incrementally from 66% at 500°C to 72% at 800°C. This shows that the majority of the chemical present are represented by diesel fuel herein. It also points towards the fact that tars are cracked within the primary reaction to produce this range of chemicals [15].

4 CONCLUSION

The thermal pyrolysis of end of life tyres was achieved and studied in the temperature range between 500 and 800°C. The yield was at a maximum at 500°C (44%) which was expected giving the material a longer residence time for complete cracking and evolution to tars vis-à-vis oils. The calorific value was increasing incrementally as a function of the reactor bed temperature reaching 42 kJ kg^{-1}. Furthermore, about 72% of the oils extracted were in the diesel fuel range. The findings herein can be of immense importance for providing a circular economy platform for countries that have a major environmental problem faced with stockpiling ELTs. The pyrolysis oil is associated with a better environmental performance when compared with combustion and incineration of to provide heat. Therefore, it is quite important to consider the fact that reducing the sulphur content and physical blending of such products can provide a platform for integration with oil and gas industries that can be of economic and environment benefit.

ACKNOWLEDGEMENTS
The author would like to thank the Kuwait Foundation for the Advancement of Sciences (KFAS) for funding and supporting this research project through the Grant for Project

EM085C (PN17-44SC-03). The author would also like to thank Kuwait Municipality (KM) for their help and support to the work conducted in this research. Gratitude is also expressed to the Kuwait Institute for Scientific Research (KISR) for internal fund acquirement and support.

REFERENCES

[1] Williams, P.T., Pyrolysis of waste tyres: A review. *Waste Management*, **33**, pp. 1714–1728, 2013.

[2] Alkhatib, R., Loubar, K., Awad, S., Mounif, E. & Tazerout, M., Effect of heating power on the scrap tires pyrolysis derived oil. *Journal of Analytical and Applied Pyrolysis*, **116**, pp. 10–17, 2015.

[3] EPA, National overview: Facts and figures on materials, wastes and recycling. https://www.epa.gov/facts-and-figures-about-materials-waste-and-recycling/national-overview-facts and figures-materials. Accessed on: 26 Mar. 2021.

[4] Adhikari, B., De, D. & Maiti, S., Reclamation and recycling of waste rubber. *Progress in Polymer Science*, **25**, pp. 909–948, 2000.

[5] Hita, I., Arabiourrutia, M., Olazar, M., Bilbao, J., Arandes, J.M. & Sánchez, P.C., Opportunities and barriers for producing high quality fuels from the pyrolysis of scrap tires. *Renewable and Sustainable Energy Reviews*, **56**, pp. 745–759, 2016.

[6] Murugan, S., Ramaswamy, M.C. & Nagarajan, G., The use of tyre pyrolysis oil in diesel engines. *Waste Management*, **28**, pp. 2743–2749, 2008.

[7] de Marco Rodriguez, I., Laresgoiti, M.F., Cabrero, M.A., Torres, A., Chomón, M.J. & Caballero, B., Pyrolysis of scrap tyres. *Fuel Processing Technology*, **72**, pp. 9–22, 2001.

[8] Gieré, R., Smith, K. & Blackford, M., Chemical composition of fuels and emissions from a coal+tire combustion experiment in a power station. *Fuel*, **85**, pp. 2278–2285, 2006.

[9] Lombardi, L., Carnevale, E. & Corti, A., A review of technologies and performances of thermal treatment systems for energy recovery from waste. *Waste Management*, **37**, pp. 26–44, 2015.

[10] Martínez, J.D., Puy, N., Murillo, R., García, T., Navarro, M.V. & Mastral, A.M., Waste tyre pyrolysis – A review. *Renewable & Sustainable Energy Reviews*, **23**, pp. 179–213, 2013.

[11] Xiao, G., Ni, M.-J., Chi, Y. & Cen, K.-F., Low-temperature gasification of waste tire in a fluidized bed. *Energy Conversion and Management*, **49**, pp. 2078–2082, 2008.

[12] Al-Salem, S.M., Valorisation of end-of-life tyres (ELTs) in a newly developed pyrolysis fixed bed batch process. *Process Safety & Environmental Protection*, **138**, pp. 167–175, 2020.

[13] Al-Salem, S.M., Lettieri, P. & Baeyens, J., Kinetics and product distribution of end of life tyres (ELTs) pyrolysis: A novel approach in polyisoprene and SBR thermal cracking. *Journal of Hazardous Materials*, **172**(2–3), pp. 1690–1694, 2009.

[14] Daily Mail, World's biggest tyre graveyard: Incredible images of Kuwaiti landfill site that is home to seven million wheels and so huge it can be seen from space, 2013. http://www.dailymail.co.uk/news/article-2337351. Accessed on: 26 Mar. 2021.

[15] Aguado, R., Arrizabalaga, A., Arabiourrutia, M., Lopez G., Bilbao, J. & Olazar, M., Principal component analysis for kinetic scheme proposal in the thermal and catalytic pyrolysis of waste tyres. *Chemical Engineering Science*, **106**, pp. 9–17, 2014.

[16] Mazloom, G., Farhadi, F. & Khorasheh, F., Kinetic modeling of pyrolysis of scrap tires. *Journal of Analytical and Applied Pyrolysis*, **84**(2), pp. 157–164, 2009.

[17] Mui, E.L.K., Lee, V.K.C., Cheung, W.H. & McKay, G., Kinetic modeling of waste tire carbonization. *Energy & Fuels*, **22**(3), pp. 1650–1657, 2008.

[18] Banar, M., Akyıldız, V., Özkan, A., Çokaygil, Z. & Onay, Ö., Characterization of pyrolytic oil obtained from pyrolysis of TDF (Tire derived fuel). *Energy Conversion and Management*, **62**, pp. 22–30, 2012.

[19] Lopez, G. et al., Waste truck-tyre processing by flash pyrolysis in aconical spouted bed reactor. *Energy Conversion and Management*, **142**, pp. 523–532, 2017.

[20] Crown Oil, Why are sulphur levels being limited? 2020. https://www.crownoiluk.com/sulphur-limits-on-fuel-explained/#:~:text=The%20 sulphur%20content%20of%20class%20D%20diesel%20is,lower%20in%20cost%20 when%20compared%20to%20road%20diesel. Accessed on: 26 Mar. 2021.

[21] Hristova, M. & Tchaoushev, S., Calculation of flash points and flammability limits of substances and mixtures. *Journal of the University of Chemical Technology and Metallurgy*, **41**, pp. 291–296, 2006.

[22] Mirmiran, S., Pakdel, H. & Roy, C., Characterization of used tire vacuum pyrolysis oil: Nitrogenous compounds from the naphtha fraction. *Journal of Analytical and Applied Pyrolysis*, **22**, pp. 205–215, 1992.

[23] Li, S.Q., Yao, Q., Chi, Y., Yan, J.H. & Cen, K.F., Pilot-scale pyrolysis of scrap tires in a continuous rotary kiln reactor. *Industrial & Engineering Chemistry Research*, **43**, pp. 5133–5145, 2004.

[24] Lopez, G., Olazar, M., Amutio, M., Aguado, R. & Bilbao, J., Influence of tire formulation on the products of continuous pyrolysis in a conical spouted bed reactor. *Energy & Fuels*, **23**, pp. 5423–5431, 2009.

PYROLYSIS OF END OF LIFE TYRES RECLAIMED FROM LORRY TRUCKS: PART II – ANALYSIS OF RECOVERED CHAR

SULTAN MAJED AL-SALEM
Environment & Life Sciences Research Centre, Kuwait Institute for Scientific Research, Kuwait

ABSTRACT

Annual production (and subsequently disposal) of tyres is taking place growing proportionally to the increase in population. Pyrolysis has proven to be a good thermos-chemical conversion method that can convert end of life tyres (ELTs) into valuable products namely solid char. In this work, experimental studies were carried out on end of life tyres (ELTs) reclaimed from lorry trucks which are known to be quite resilient to environmental exposure. The cracking took place between 500 and 800°C in a pyrolysis operation of fixed bed reactor type. The char recovered was extensively analysed in this study. Average particle size analysis showed that majority of the recovered char was >4 mm (36.2%) and the moisture content was somewhat low around 1 wt.% determined by thermogravimetry. On the other hand, the gross calorific value (GCV) was determined to be higher than 28 kJ·g^{-1} pointing towards possibility of using this product as a source of energy.
Keywords: tyres, pyrolysis, diesel, energy, waste.

1 INTRODUCTION

End of life tyres (ELTs) represent a major component of solid waste (SW) that requires special attention. This is due to the fact that ELTs don't degrade with environmental exposure and can cause major environmental pollution when stockpiled. Various countries around the world have started implementing regulations and enforce environmental legislations to protect their environments from ELTs accumulation. In a European context, the *Vehicle Life Cycle End* Directive enforces 40% (as a minimum) of ELTs to be valorised and redirected from landfill sites [1].

ELTs are composed of various elastomers and rubbers that are typically bound together with carbon black fillers [2]. Evans and Evans [3] have shown that passenger vehicles and lorry truck tyres contain the following in their composition (wt.%): Rubber (45–47%), carbon black (21.5–22%), metal (16.5–21.5%), textiles (5.5%), zinc oxide (1–2%), sulphur (1%) and additives (5–7.5%). The composition of ELTs encourages the valorisation of the rubber constituents to produce value added products such as fuel oil and gases. To do so, pyrolysis presents itself as a solution that can reduce the volume of tyre waste and extract valuable products from it. Pyrolysis is defined as a thermo-chemical conversion method that can treat feedstock using inert atmospheres in a temperature range between 350 and 900°C [4]. Pyrolysis produces fuel in the form of oil that can be used as an added product to refinery blends or directly as a fuel. Gases from ELTs pyrolysis can be used as a heating gas to the process itself since it comprises C_1 to C_4 chemicals. Furthermore, the third and last product of the process (i.e., solid char) can be used as a solid fuel or upgraded as activated carbon [2].

The yield of pyrolysis char varies significantly with respect to the type of reactor used and operating conditions. Fixed bed reactors implementing a temperature range between 300 and 1,000°C have been reported to produced pyrolysis char from ELTs in the range from 30 to ≈48 wt.% [5]–[12]. On the other hand, fluidised bed reactors (FBRs) have been reported to produce char in the range from 35 to 48.5 wt.% in temperatures that range

WIT Transactions on Engineering Sciences, Vol 133, © 2021 WIT Press
www.witpress.com, ISSN 1743-3533 (on-line)
doi:10.2495/MC210121

between 450 and 780°C [13], [14]. In this work, lorry (truck) ELT grade was studied using a fixed bed reactor to achieve pyrolysis conditions with the aim of studying the pyrolysis char (pyro-char) quality. This part of the communication compliments the first part which reports the oil properties extracted from the same experimental set using the same feedstock. The work presented herein can also pave the way for strategies in Middle-East and by extension other regions to have a circular economy using ELTs as a potential feedstock material with value added products recovered.

2 EXPERIMENTAL

ELTs carcasses (10 kg) were acquired from Al-Essa Company (Kuwait) which were reclaimed ensuring similar make and model of each tyre grade. The ELTs were of lorry truck type reclaimed originally from Al-Maillam group dealership (315/80R22.5). Each tyre was firstly air blown to remove any dust particles and then subjected to shredding using ELDAN (Denmark) at 60°C. The obtained samples (manually measured) were 1.22 cm in size. The samples were stored in laboratory conditions ($\approx$ 22°C) using sealed plastic containers. Feedstock in the amount of 200 g was placed in the fixed bed as a charge with Alumina packing of a 5 mm diameter (average bulk density of 700 kg·m^{-3}) was used in the amount of 120 g. For operation and reactor specifications, the readers are referred to Al-Salem [4].

Char yield was studied between 500 and 800°C as an average of the three bed temperatures used in the operation. A Tarsus F3 TGA from NETZSCH was used for the thermal decomposition properties of the samples and for the determination of the moisture content. The TGA unit is externally calibrated and the analysis was conducted as per ISO 11358 directions using a 10 mg sample weight. A CAL3K Advanced bomb calorimeter was used for gross calorific value (CV) determination of the oil samples as per ISO 1928:20 using 0.5 g specimens. Average particle size was measured on the char samples as they were supplied using the sieve analysis method. Sieves of mesh sizes 10 mm, 4 mm, 2 mm, 1.60 mm, 1 mm, 0.850 mm and 0.125 mm were stacked in order of highest to the lowest size and assembled on a sieve shaker A sample mass of about 90 g was taken from each char sample and poured on the topmost sieve (mesh size 10 mm).

3 RESULTS AND DISCUSSION

Fig. 1 shows the char yield as a function of the reactor bed temperature. The amount of char reduced from 91 to 85 g between 500 and 800°C, respectively. Lower residence time in the pyrolysis reactor can influence such a behaviour. The higher the final temperature the longer the material will degrade. The char represents some 45% of the total charge of the feedstock which falls in the range previously mentioned in the introduction section of past reports. The moisture content ranged between 0% and 2.16% depending on the pyrolysis temperature used to produce the char with an average of all samples around 1 wt.%. Past reports showed that pyrolysis char had a moisture content as high as 3.57 wt.% [2]. The calorific value was estimated for the char and is depicted in Fig. 2.

The char consists mainly of the original carbon black present in the feedstock [15]. Char derived from pyrolysis is considered a primary reaction product, given that it accounts for 30–40% of the original tyre mass and is mainly made up of carbon black and a significant quantity of inorganic ash containing most of the sulphur and Zn used in the tyre manufacturing [16]–[19]. Antoniou and Zabaniotou [20] showed that the higher heating value (HHV) of the char derived from ELTs pyrolysis between 450 and 600°C was 29.75 MJ/kg which made them determine it as a good fuel for potential upgrading. All

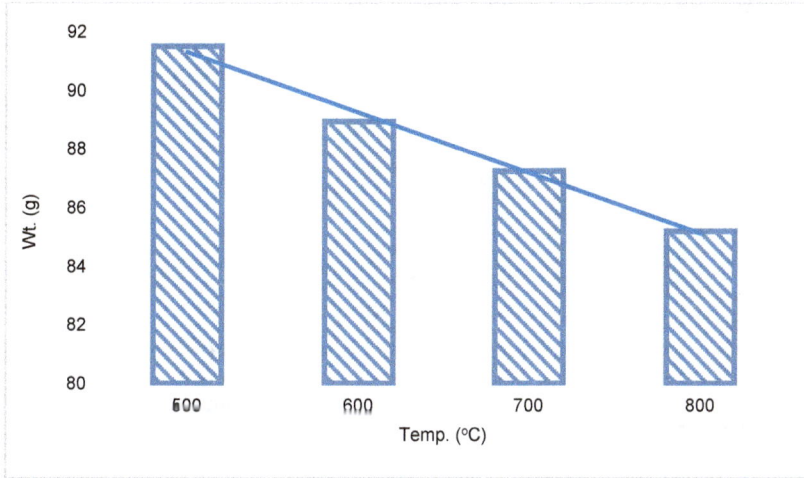

Figure 1: Pyrolysis char calorific yield (g) as a function of the operating temperature (°C).

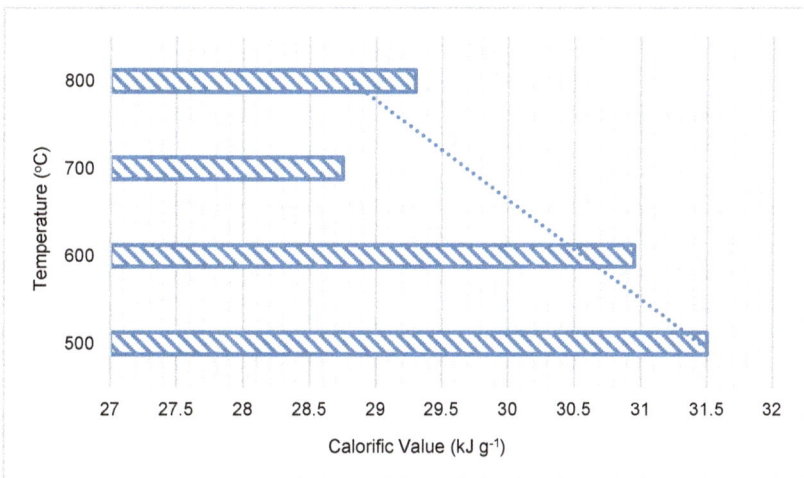

Figure 2:   Pyrolysis char calorific value (kJ g^{-1}) as a function of the operating temperature (°C).

results in this work determining heating value were higher than the aforementioned ranged which promotes the pyro-char as a god candidate for fuel use. Therefore, it is recommended to pyrolyse tyres in Kuwait for potential char as pyro-char fuel. Antoniou and Zabaniotou [20] have also shown that char was at large a macroporous material (>50 nm, IUPAC classification). Based on the above, ELT char can be used as a solid fuel, or as a precursor for porous material production [21]. Further upgrading could be achieved for the char (to eliminate ash and residues) as well via acidification/basification treatments [21]. This is due to the fact that majority of the char analysed is larger in size than 4 and 10 mm (Fig. 3).

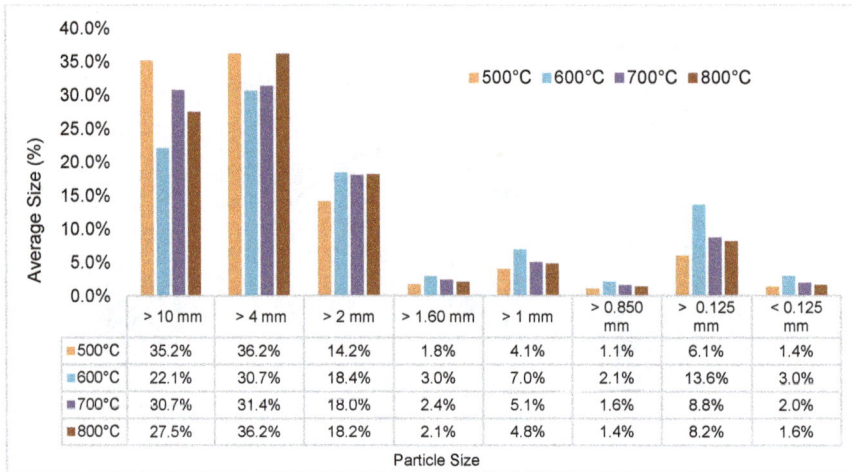

	> 10 mm	> 4 mm	> 2 mm	> 1.60 mm	> 1 mm	> 0.850 mm	> 0.125 mm	< 0.125 mm
500°C	35.2%	36.2%	14.2%	1.8%	4.1%	1.1%	6.1%	1.4%
600°C	22.1%	30.7%	18.4%	3.0%	7.0%	2.1%	13.6%	3.0%
700°C	30.7%	31.4%	18.0%	2.4%	5.1%	1.6%	8.8%	2.0%
800°C	27.5%	36.2%	18.2%	2.1%	4.8%	1.4%	8.2%	1.6%

Figure 3:   Average (%) and particle size and as a function of the operating temperature (°C).

4 CONCLUSION AND FUTURE WORK

The pyrolysis char investigated in this work showed potential for future use as a solid fuel source of a size that exceeds macroporous materials with a calorific value ranging between 29.3 and 31.2 kJ·g^{-1}. To fully investigate the potential of this material, future analysis at this stage is required. Firstly, the sulphur content of the material should be assessed. This will indicate the potential of using it as a potential environmental source of energy in the future. The density of the char should also be considered in the future to compare it with other common solid fuels (e.g., refuse derived fuel). In addition, there should be a somewhat comparable database possessed with governments and interested parties showing the potential of products from various feedstock use (ELTs). This will truly provide a platform for future use in integration platforms with oil and gas industries to reduce fossil fuel dependence.

ACKNOWLEDGEMENTS

The author would like to thank the Kuwait Foundation for the Advancement of Sciences (KFAS) for funding and supporting this research project through the Grant for Project EM085C (PN17-44SC-03). The author would also like to thank Kuwait Municipality (KM) for their help and support to the work conducted in this research. Gratitude is also expressed to the Kuwait Institute for Scientific Research (KISR) for internal fund acquirement and support.

REFERENCES

[1]   Hita, I., Arabiourrutia, M., Olazar, M., Bilbao, J., Arandes, J.M. & Sánchez, P.C., Opportunities and barriers for producing high quality fuels from the pyrolysis of scrap tires. *Renewable and Sustainable Energy Reviews*, **56**, pp. 745–759, 2016.

[2]   Williams, P.T., Pyrolysis of waste tyres: A review. *Waste Management*, **33**, pp. 1714–1728, 2013.

[3]   Evans, A. & Evans, R., The composition of a tyre: Typical components, *Waste & Resources Action Programme*, Banbury, UK, 2006.

[4] Al-Salem, S.M., Valorisation of end of life tyres (ELTs) in a newly developed pyrolysis fixed bed batch process. *Process Safety & Environmental Protection*, **138**, pp. 167–175, 2020.

[5] Aydın, H. & Ilkılıc, C., Optimization of fuel production from waste vehicle tires by pyrolysis and resembling to diesel fuel by various desulfurization methods. *Fuel*, **102**, pp. 605–612, 2012.

[6] Leung, D.Y.C., Yin, X.L., Zhao, Z.L., Xu, B.Y. & Chen, Y., Pyrolysis of tire powder: Influence of operation variables on the composition and yields of gaseous product. *Fuel Processing Technology*, **79**, pp. 141–155, 2002.

[7] Miranda, M., Pinto, F., Gulyurtlu, I. & Cabrita, I., Pyrolysis of rubber tyre wastes: A kinetic study. *Fuel*, **103**, pp. 542–552, 2013.

[8] Williams, P.T., Besler, S. & Taylor, D.T., The pyrolysis of scrap automotive tyres: The influence of temperature and heating rate on product composition. *Fuel*, **69**, pp. 1474–1482, 1990.

[9] Cunliffe, A.M. & Williams, P.T., Composition of oils derived from the batch pyrolysis of tyres'. *Journal of Analytical and Applied Pyrolysis*, **44**, pp. 131–152, 1998.

[10] Williams, P.T., Bottrill, R.P. & Cunliffe, A.M., Combustion of tyre pyrolysis oil. *Transactions of the Institution of Chemical Engineers*, **76**, pp. 291–301, 1998.

[11] Banar, M., Akyıldız, V., Ozkan, A., Cokaygil, Z. & Onay, O., Characterization of pyrolytic oil obtained from pyrolysis of TDF (Tire derived fuel). *Energy Conversion and Management*, **62**, pp. 22–30, 2012.

[12] Laresgoiti, M.F., Caballero, B.M., de Marco, I., Torres, A., Cabrero, M.A. & Chomón, M.J., Characterization of the liquid products obtained in tyre pyrolysis. *Journal of Analytical and Applied Pyrolysis*, **71**, pp. 917–934, 2004.

[13] Kaminsky, W., Mennerich, C. & Zhang, Z., Feedstock recycling of synthetic and natural rubber by pyrolysis in a fluidized bed. *Journal of Analytical and Applied Pyrolysis*, **85**, pp. 334–337, 2009.

[14] Williams, P.T. & Brindle, A.J., Fluidised bed pyrolysis and catalytic pyrolysis of scrap tyres. *Environmental Technology*, **24**, pp. 921–929, 2003.

[15] Aguado, R., Arrizabalaga, A., Arabiourrutia, M., Lopez, G., Bilbao, J. & Olazar, M., Principal component analysis for kinetic scheme proposal in the thermal and catalytic pyrolysis of waste tyres. *Chemical Engineering Science*, **106**, pp. 9–17, 2014.

[16] Lopez, F.A., Centeno, T.A., Alguacil, F.J. & Lobato, B., Distillation of granulated scrap tires in a pilot plant. *Journal of Hazardous Materials*, **190**, pp. 285–292, 2011.

[17] Lopez, G., Aguado, R., Olazar, M., Arabiourrutia, M. & Bilbao, J., Kinetics of scrap tyre pyrolysis under vacuum conditions. *Waste Management*, **29**, pp. 2649–2655, 2009.

[18] Lopez, G., Olazar, M., Aguado, R. & Bilbao, J., Continuous pyrolysis of waste tyres in a conical spouted bed reactor. *Fuel*, **89**, pp. 1946–1952, 2010.

[19] Lopez, G., Olazar, M., Amutio, M., Aguado, R. & Bilbao, J., Influence of tire formulation on the products of continuous pyrolysis in a conical spouted bed reactor. *Energy & Fuels*, **23**, pp. 5423–5431, 2009.

[20] Antoniou, N. & Zabaniotou, A., Experimental proof of concept for a sustainable end of life tyres pyrolysis with energy and porous materials production. *Journal of Cleaner Production*, **101**, pp. 323–336, 2015.

[21] Danon, B., Mkhize, N.M., van der Gryp, P. & Görgens, J.F., Combined model-free and model-based devolatilisation kinetics of tyre rubbers. *Thermochimica Acta*, **601**, pp. 45–53, 2015.

SECTION 3
EMERGING AND
GREEN MATERIALS

ENHANCED PHYSICAL PROPERTIES OF NANOCELLULOSE FIBER-REINFORCED GREEN COMPOSITES

HITOSHI TAKAGI, HIROAKI GENTA & ANTONIO NORIO NAKAGAITO
Tokushima University, Japan

ABSTRACT

As a potential nanoscale reinforcement, nanocellulose fibers (NCFs) have drawn much scholarly attention. The NCFs whose typical diameter is 5–100 nm exhibited potentially high mechanical properties, and its tensile strength and modulus were estimated to be 1.7 GPa and 140 GPa. These higher tensile strength of the NCF is reported to be equivalent to those of glass fiber. However, the resultant mechanical properties reported for the NCF reinforced composites are lower than estimated values. In this study, we tried to enhance their mechanical properties by applying the following two approaches. These approaches are optimization in the fibrillation condition of NCFs by using grinding treatment and ultrasonic treatment. The effectiveness of the proposed two approaches has been successfully demonstrated experimentally; tensile strength and Young's modulus of the NCF-reinforced composites ultra-sonicated for 60 min were improved by 70% and 55%, respectively, compared to that of the untreated composites.
Keywords: nanocellulose fiber, green composites, polyvinyl alcohol, dispersion.

1 INTRODUCTION

At present, many kinds of plastic products are used in our daily life. Among these plastic products, fiber-reinforced plastics (FRPs) are used as structural members in automobiles, aircraft, etc. However, the FRPs are generally non-biodegradable and thus difficult to dispose of. For this reason, FRPs have the disadvantage of having a significant impact on the environment when disposed of. In order to obtain alternative composite materials with low environmental impact, research and development of green composites using biodegradable resin as a base matrix material and natural fiber as a reinforcing phase are being conducted continuously [1]–[3].

In this study, we focused on green composites reinforced with nanocellulose fiber (NCF), which is abundant on the earth and has high strength. In general, it is known that the properties of nanocomposites vary not only with the individual properties of the matrix and the nanoscale reinforcement but also with various factors such as interfacial adhesion, uniform distribution, and orientation of the reinforcement. However, there are many unclear points about its concrete contribution to the mechanical performances of the nanocomposites.

In the past, there have been studies on the preparation and characterization of nanocomposite materials combining the biodegradable polylactic acid (PLA) resin and NCF [4]. This study reported that it is difficult to fabricate nanocomposite material at the nano-order level in the PLA resin because strong cohesive force acts among the NCF and the reinforcing effect of NCF was not sufficiently obtained. To overcome this difficulty in the dispersion of NCF, they are studying a new dispersing technology using melt kneading. They confirmed that composites with aggregates of NCF of several tens of μm had improved strength compared to those with aggregates of NCF of several hundred μm. They also reported that the uniform dispersion of finer reinforcement in the resin matrix greatly contributed to the improvement in the strength of NCF-reinforced composite materials.

WIT Transactions on Engineering Sciences, Vol 133, © 2021 WIT Press
www.witpress.com, ISSN 1743-3533 (on-line)
doi:10.2495/MC210131

However, the details of the effect of the size and dispersion of NFC is still not elucidated. In this study, we used polyvinyl alcohol (PVA)/NCF composite as a model material and focused on uniform dispersion of NCF reinforcement in PVA matrix and aimed at the development of high-strength NCF-reinforced polymer composites.

In this study, we used PVA/NCF nanocomposite as a model material. We focused on the effect of uniform dispersion of the NCF reinforcement in the PVA matrix and orientation of the reinforcement and aimed at the development of high-strength nanocomposite.

2 EXPERIMENTAL METHOD

2.1 Materials

In this study, we used Celish (KY-100G, Daicel Co., Japan) as NCF. The Celish is commercially available NCF in which wood pulp is microfibrillated to nanoscale cellulose fibrils by high-pressure homogenization treatment. PVA, a water-soluble biodegradable polymer (162–16325, Wako Pure Chemical Industries, Ltd., Japan) was used as a matrix polymer. PVA has excellent water-solubility and has characteristics that hydrophilic reinforcing phase such as NCF can be easily dispersed in the resin.

2.2 Preparation of preform sheets

First, the PVA powder of 25 g was poured into room temperature water of 475 g with stirring. The mixture (PVA content = 5 wt%) was then heated and dissolved using a mantle heater (HB-1000T, As One Corporation, Japan) with stirring. The Celish was in semi-solid form and the NCFs are highly agglomerated, so it was stirred with distilled water for 24 hours to prepare a 1 wt% NCF suspension. The aqueous PVA solution and the NCF suspension were mixed and stirred for another 24 hours. The resultant mixed suspension was cast into a polystyrene container and dried in a drying oven at 30°C for 24 hours to prepare PVA/NCF preform sheet, and then a further 100 g of the mixture solution was added on the PVA/NCF preform sheet which was not completely dried. This treatment was repeated six times to produce a thick preformed sheet. The final NCF content in the PVA/NCF composite sheet was 30 wt%.

2.3 Grinding treatment

The mixed suspension produced (NCF content = 1 wt%) was passed through a grinder (Supermasscolloider MKCA6-2, Masuko Sangyo Co., Ltd., Japan, Fig. 1) with a rotation speed of 1,500 rpm to refine NCF. This grinding process was performed up to twice.

2.4 Ultrasonication treatment

An ultrasonic oscillator (UH-150, SMT Co. Ltd., Japan, Fig. 2) was used to microfibrillate the NCF while stirring the PVA/NCF mixture (NCF content = 30 wt%) with a stirrer. A set of this microfibrillation treatment was ultrasonication treatment for 15 min. and cooling for 15 min. and this treatment was repeated up to four times (total ultrasonication time was 60 min.).

Figure 1: A grinder used in this study [5].

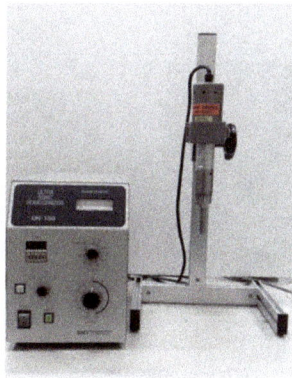

Figure 2: An ultrasonication machine used in this study.

3 RESULTS AND DISCUSSION

The results of the tensile test for the PVA/NCF nanocomposites are shown in Fig. 3. It can be seen that both the tensile strength and Young's modulus increase as the number of grinding treatment increases. Comparing the results of untreated specimens and specimens with twice grinding treatment, the tensile strength and Young's modulus of the treated specimen are improved by 43% and 28%, respectively.

Fig. 4 shows the results of SEM observation of the degree of fibrillation of NCF in order to investigate the effects of grinding treatment. We can see that some thick NCFs still remain in the untreated condition, on the other hand much finer NCFs are produced after grinding treatment. It should be noted that almost no thick NCF exists after twice grinding treatment. Similar experimental results regarding such refinement were also reported elsewhere [5].

Fig. 5 depicts the effect of ultrasonication treatment on the tensile properties of PVA/NCF nanocomposites. Both tensile strength and Young's modulus increase with increasing ultrasonication time. The highest tensile properties are obtained in the PVA/NCF nanocomposites ultrasonicated for 60 min. As compared with the untreated PVA/NCF

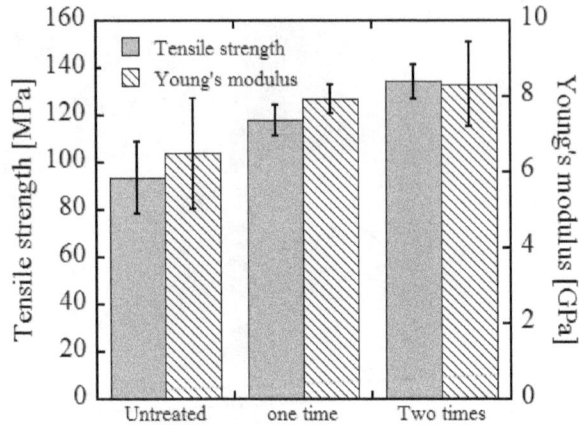

Figure 3: Tensile properties of the nanocomposites after grinding treatment.

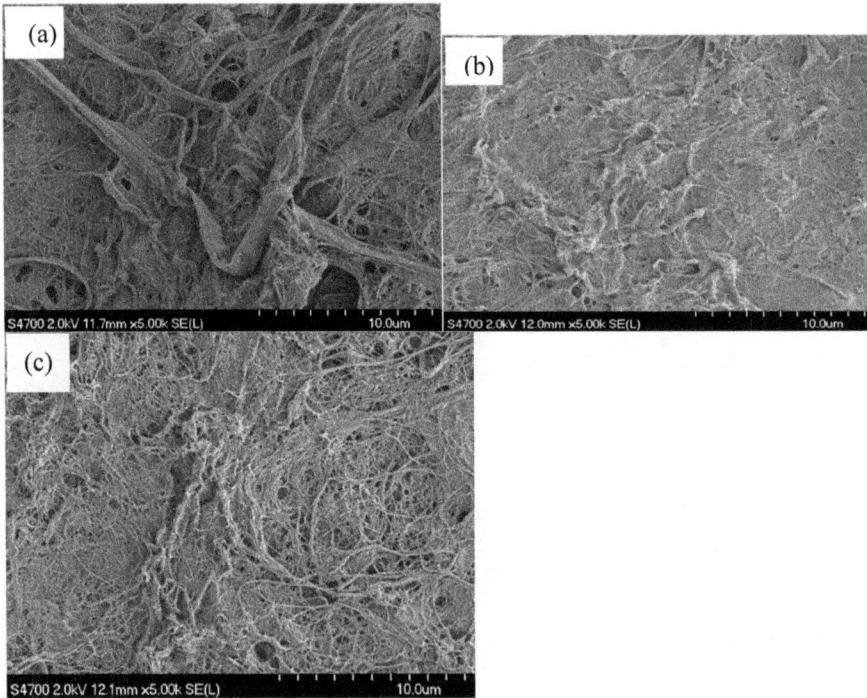

Figure 4: SEM photomicrographs of (a) Untreated NCF and grinding treatment of (b) One time and (c) Two times.

nanocomposites, the tensile strength and Young's modulus of the nanocomposites ultra-sonicated for 60 min. are improved by 70% and 55%, respectively. The strength level of the nanocomposites ultrasonicated for 60 min was comparable to that of mechanical extension-treated nanocomposites with 30 wt% NCF [6].

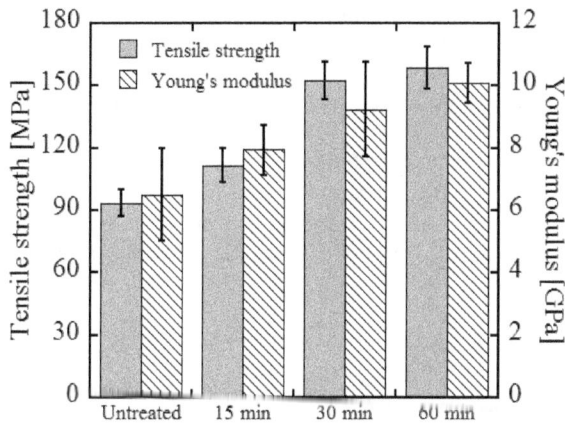

Figure 5: Tensile properties of the nanocomposites after ultrasonication treatment.

As in the case of the grinding treatment, the NCF subjected to the ultrasonic treatment has a smaller thick fiber instead. It can be seen that thick cellulose fibers are reduced in the NCF after the ultrasonic treatment as in the case of the grinding treatment. The following are conceivable causes of the improvement of the strength characteristics accompanying the finer fibers. As the fibers are finely fibrillated, the surface area of the fibers increases. Therefore, the bonding area at the interface between the resin matrix and NCF is increased, and the interfacial region where hydrogen bonding occurs is increased so that the strength is improved.

4 CONCLUSIONS

By performing the grinding treatment and the ultrasonic treatment, the tensile strength and Young's modulus of the PVA/NCF nanocomposite material are improved. It was suggested that the cause of the improvement in the mechanical properties of the PVA/NCF nanocomposites was related to the refinement of the NCF and the improvement of the uniform dispersion of the NCF in the PVA matrix.

ACKNOWLEDGEMENT

We acknowledge financial support from JSPS, KAKENHI project number 16H01790.

REFERENCES

[1] Netravali, A.N. & Chabba, S., Composites get greener. *Materials Today*, 6(4), pp. 22–29, 2003.
[2] Takagi, H., Review of functional properties of natural fiber-reinforced polymer composites – Thermal insulation, biodegradation and vibration damping properties. *Advanced Composite Materials*, **28**(5), pp. 525–543, 2019.
[3] Katogi, H. & Takemura, K., Flexural properties of flax sliver reinforced green composite by molding pressure and chitosan fiber addition. *WIT Transactions on Engineering Sciences*, vol. 124, WIT Press: Southampton and Boston, pp. 93–99, 2019.

[4] Okada, K. & Uemoto, S., Properties of PLA composites which filled microfibrillated cellulose fiber (MFC). *Journal of the Textile Machinery Society of Japan*, **61**(9), pp. 621–626, 2008 (in Japanese).
[5] Takagi, H., Nakagaito, A.N. & Sakaguchi Y., Structural modification of cellulose nanocomposites by stretching. *WIT Transactions on Engineering Sciences*, vol. 116, WIT Press: Southampton and Boston, pp. 251–256, 2017.
[6] Takagi, H., Nakagaito, A.N., Nishimura, K. & Matsui, T., Mechanical characterisation of nanocellulose composites after structural modification. *WIT Transactions on the Built Environment*, vol. 166, WIT Press: Southampton and Boston, pp. 335–341, 2016.

COMPARISON OF BIOMASSES AS ADSORBENT MATERIALS FOR PHENOL REMOVAL

PUSHPA JHA
Department of Chemical Engineering, Sant Longowal Institute of Engineering & Technology, India

ABSTRACT

India is a producer of a colossal number of biomasses with high quantity. Even after using them for energy generation, large proportions of residues remain unutilised. They could be utilised as an adsorbent-material to get rid of phenol from aqueous streams. Phenol is listed as highly toxic as per available databases. Thermo-chemical treatment methods have been widely reported to improve the characteristics of biomass-based adsorbents. In this work, based on the availability, three biomasses, Acacia Nilotica Branches (AC), Lantana Camera (LA) and Rice-Husk (RI), were given the treatment. The resulting activated forms of adsorbents were named activated Acacia Nilotica Branches (ACC), activated Lantana Camera (LAC) and activated Rice Husk (RIC). The materials obtained had a high content of fixed carbon, iodine number, BET surface area, and methylene blue adsorption. The operating parameters for sorption in terms of dosage, pH, time of contact, initial phenol concentration and agitation speed were optimised. At these conditions, the adsorption isotherms were compared, and they were explained by Langmuir, Freundlich, and Temkin models. LAC and RIC, respectively highest, followed sorption capacity of ACC. Kinetics of the process on adsorbents considered followed pseudo-first-order and pseudo-second-order models.

Keywords: acacia nilotica branches, adsorbent, adsorption isotherms, adsorption kinetics, adsorption parameters, biomass, characterisation, Lantana camera, rice-husk, phenol.

1 INTRODUCTION

A large number of biomasses are produced every year [1], [2]. Farmers are seen burning them as a solution to clearing and preparing the fields for the next crop. The process generates greenhouse gases. Biomasses have a high moisture content and low bulk density, making them difficult to transport from one place to another. So it is mandatory to investigate various possible ways to utilise them at the exact location as its production [3].

The literature review gives an account of various biomasses applications as adsorbents for the removal of organic compounds, especially phenols in industrial effluents [4]–[8]. In this work, three biomasses, Acacia Nilotica Branches (AC), Lantana Camera (LA), and Rice Husk (RI), has been considered to be studied (based on their availability) for the removal of phenol. The presence of phenol in water streams affects the food chain. The pollutant finds its way to water sources mainly through the effluents from paint, pesticides, coal conversion, polymeric resin, petroleum, and petrochemicals industries [9]. Phenol is listed as a high priority chemical in various databases [10].

In this work, AN, LA and RI were selected based on their suitability [11] and availability in the region. The author's work establishes that the maximum removal of phenol was no more than 35% when the biomasses were used as adsorbent without activation. Properties of AN, LA and RI were compared with commercial-grade carbon. It is observed that the biomasses have much lower BET surface areas, lower fixed carbons (FC) and higher volatile matter (VM). Therefore to upgrade its adsorbent properties, they were given thermochemical treatment based on a literature review [3]. The treatment helped to lower VM and ash percentages, thereby increasing the FC percentages. Consequently, BET surface areas and methylene blue adsorption values of the adsorbents increased. Therefore, activated forms of AN, LA, and RI were abbreviated as ACC, LAC, and RIC.

WIT Transactions on Engineering Sciences, Vol 133, © 2021 WIT Press
www.witpress.com, ISSN 1743-3533 (on-line)
doi:10.2495/MC210141

The adsorption's operating parameters were effects of dosage, pH, time of contact, initial phenol concentration, and agitation speed. They were optimised for the generation of adsorption curves that fit into Langmuir, Freundlich and Temkin isotherms. The kinetics of adsorption was studied, and they were better explained by pseudo-first-order and pseudo-second-order kinetics. Adsorption of phenol on ACC, LAC and RIC could be studied in the supernatant concentration range of 100 to 975 mg/l.

2 MATERIALS AND METHODS

Various experiments were conducted for the present work. Details are mentioned in the following subsections.

2.1 Preliminary study of biomasses

Biomasses were analysed in bulk densities and proximate analysis as per standard methods and ASTM: D3173-75. The testing determined moisture, VM, FC and ash percentages of biomasses selected [12].

2.2 Preparation of adsorbents

Biomasses AN, LA and RI were separately acquired, washed and ground in powder form, which passes a 300-micron screen. The samples were treated with 30% H_3PO_4 for 3 hours at 37°C. The resulting slurry was filtered, and the residue was washed thoroughly. The residues obtained were dried and pyrolysed at 500°C in the muffle furnace. The char was digested with NaOH solution (12%) for 5 hours at 68°C. The resulting solid particles were filtered out, washed and dried. The samples thus obtained were powdered and stored for further studies [12].

2.3 Characterisation of adsorbents

The stored powder of adsorbents (ACC, LAC and RIC) were subject to individual determinations of phenol number, pH, iodine number, methylene blue number, BET surface areas and particle size using standard procedures [12].

2.4 Adsorption experiments on activated biomass materials

For adsorption experiments, the activated adsorbents were taken in 100 ml of phenol solution (1 g/l) at 25°C. The parameters chosen to affect adsorption were dosage, pH, contact time, initial phenol concentration and agitation speed. The phenol concentration was measured using a double beam spectrophotometer. The equilibrium was studied by stirring the solution for 24 hours at 200 pm at the temperature of 25°C. Solutions were maintained at pH ranging from 2.0 to 12.0 with the help of 0.1NaOH and $0.1H_2SO_4$. The adsorption parameters were optimised, which were used to study the effect of time on the process. The extent of adsorption was measured at a different time until constancy in phenol removal was ascertained [12].

3 RESULTS AND DISCUSSIONS

Results of various experiments performed, as mentioned in the previous sections, are discussed under various subheadings as follows.

3.1 Preliminary study of biomasses

Bulk densities of biomasses and proximate analysis were determined on AN, LA and RI as explained in Section 2.1. The bulk density of AN and LA were above 200.0 kg/m^3 (Table 1), leading to the fact that they could be economically transported with ease. However, the identical value for RI is 110.0 kg/m^3, meaning; it needs briquetting. All three biomasses considered had a considerably high FC percentage, a desired characteristic of adsorbents for phenol transfer. Also, VM and Ash percentages of all biomasses (except the ash content of 2.5% of AN) were high, making them all the more suitable to be considered adsorbents [13].

Table 1: Preliminary study of biomasses.

Biomass	FC (%)	Ash (%)	VM (%)	Bulk density (kg/m^3)
AN	25	2.50	72.50	210.0
LA	23.50	12.30	64.20	250.0
RI	20.00	18.00	62.00	110.0

3.2 Characterisation of adsorbents

As per earlier studies by the author, selected biomasses without activation had feeble adsorption capacities. Commercial-grade carbon (CGC) has an excellent affinity for phenol. It could be because CGC generally has a high FC value, BET surface area, iodine number, and methylene blue adsorption. It has low values of ash and VM [14]. As mentioned in the Materials and Methods section, thermochemical treatment was given to all the biomasses. After comparison, ACC was the best adsorbent (Table 2). It has BET surface area comparable to CGC [15].

Table 2: Characterisation of the activated biomasses chosen as adsorbents [12].

Properties	ACC	LAC	RIC
Ash content on a dry basis (%)	1.0	5.0	4.80
FC on a dry basis (%)	87.0	85.0	84.20
VM on a dry basis (%)	12.0	10.0	11.00
pH of slurry	6.5 (1.2%)	7.5 (1.0%)	6.8 (1.1%)
Phenol number (g)	0.8	1.0	–
Iodine number	870	325	750
BET surface area (m^{2}g^{-1})	450	151	301
Particle size (μm)	45	42.3	25.3
Methylene blue adsorption (mg/g)	155	50.1	55.5

3.3 Adsorption experiments on activated biomass materials

The parameters affecting adsorption, namely, adsorbent dosage, initial phenol concentration, agitation speed, contact time, and pH of the adsorbate, were optimised. Fig. 1 indicates that for ACC, LAC and RIC, the removal of phenol increased with dosages. The maximum removal of phenol was possible with 1.2 g of ACC, 1 g of LAC and 1.1 g of RIC.

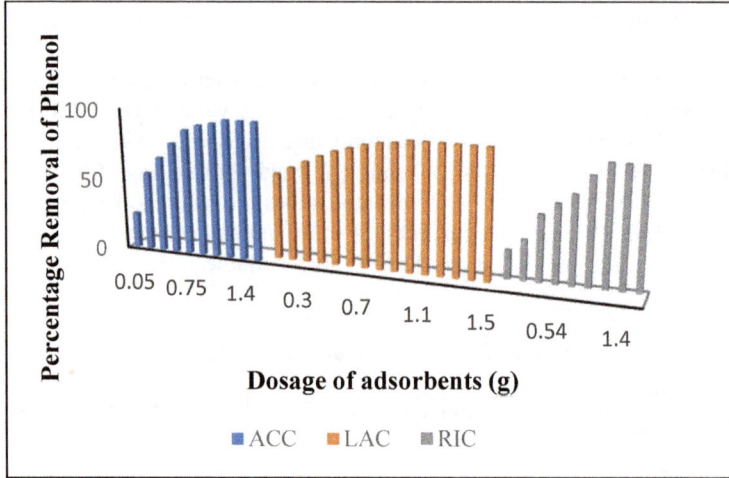

Figure 1: Variation of percentage phenol removal with the dosage of adsorbents.

As pH is one of the parameters for adsorption, its effect was measured. ACC, LAC and RIC best adsorbed the phenol at pH of 6.5, 7.5 and 6.8, respectively. Initial concentrations of phenol in the solution did not affect adsorption [12]. Agitation speed of 200 rpm in the adsorbate-adsorbent system caused maximum removal of phenol at 97%, 90% and 83% by ACC, LAC and RIC, respectively. Speed beyond 200 rpm did not affect the adsorption process.

Time is one of the most critical factors for any adsorption to take place [6]. The parameter was studied at optimised dosage conditions of the adsorbents, pH, initial phenol concentrations, agitation speeds and contact times (Section 2.4). ACC, LAC and RIC removed maximum phenol with a contact time of 4, 5 and 3.5 h correspondingly (Fig. 2).

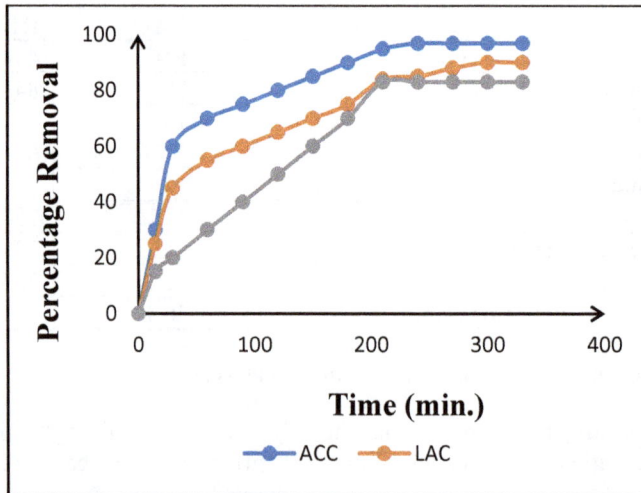

Figure 2: The effect of the contact times on removal of phenol by adsorbents.

3.4 Study of adsorption isotherms of phenol on adsorbents

As per a thorough study of the effects of various parameters on phenol's adsorption on ACC, LAC, and RIC, adsorption isotherms were generated at optimised conditions. For ACC, the isotherm was obtained at a pH of 6.5 and the time of contact of 4 h. For LAC, the equilibrium study was conducted at a pH of 7.5 and the time of contact of 5 h. For RIC, the same was conducted at a pH of 6.8 and the time of contact of 3.5 h. The adsorbate-adsorbent slurries of 100 ml were stirred at 200 rpm at 25°C, for all three cases. The adsorption equilibrium data generated were matched with Langmuir, Freundlich and Temkin isotherms [12], [16]. The results are shown in Table 3.

It was possible to get adsorption isotherm in the concentration range (C_e) of 903 to 975 mg/l for ACC, 100 to 400 mg/l for LAC, 170 to 800 mg/l for RIC. The maximum adsorptions of phenol were 500 mg/g for ACC, 600 mg/g for LAC and 132 mg/g for RIC.

Table 3 confirms that ACC, LAC and RIC follow Temkin, Langmuir and Freundlich isotherms, respectively. There is monolayer adsorption, and the surface is uniform for LAC. Values of k and n for RIC indicate that phenol has strongly attracted towards the adsorbent. The surface of the RIC is heterogeneous [16], [17].

Table 3: Isotherm constants of phenol adsorbed on adsorbents.

Biomasses	Equilibrium constants									
	Langmuir				Freundlich			Temkin		
	Q_0	b	R_L	R^2	k	n	R^2	A	b_1	R^2
ACC	–	–	0.067	–	–	–	–	0.001	0.82	0.99
LAC	16.49	13.95	1.0	0.95	0.31	0.84	0.90	–	–	–
RIC	135.13	0.007	–	0.88	13.52	3.06	0.95	0.05	76	0.90

3.5 Study of adsorption kinetics of phenol on biomasses

The study is significant to know the effectiveness of adsorption. The kinetic data for adsorption were retrieved from Fig. 2. When the values were compared with the first order and second-order kinetics, it was observed that they all fit first-order kinetics closely [17]. The adsorbents ACC and LAC follow second-order kinetics also (Table 4).

Table 4: Kinetic constants of phenol adsorbed on adsorbents.

Biomasses	Kinetic constants					
	First-order kinetics			Second-order kinetics		
	q_e	k_1	R^2	q_e	h	R^2
ACC	68.93	0.015	0.93	90.91	2.50	0.98
LAC	85.43	0.012	0.94	103.09	1.84	0.99
RIC	78.72	0.009	0.96	–	–	–

4 CONCLUSIONS

Biomasses, AN, LA and RI selected for the comparison were activated using the thermochemical route resulting in ACC, LAC and RIC as adsorbents. They were compared for their physical, chemical and adsorbent properties. Ash content reduced to 1.0% for ACC, whereas LAC and RIC are 5.0% and 4.80%, respectively. FC for all activated biomasses is high and comparable at 87.0%, 85%, and 84.2%. ACC, LAC and RIC confirmed high BET surface areas comparable to CGC. The maximum removal of phenol by ACC, LAC, and RIC was 97%, 90%, and 83% individually. Factors affecting the adsorption process were optimised. RIC got to fit into first-order-kinetics, whereas ACC and LAC followed second-order kinetics. Thus biomasses AN, LA and RI after activation proved as promising adsorbent materials for removal of phenol. They have opened the possibility for their utilisation as adsorbents to remove other organic pollutants from the industrial effluents.

ACKNOWLEDGEMENT

The help of technicians and students of the Department of Chemical Engineering of SLIET for conducting various experiments required for the work presented in this paper are appreciated.

REFERENCES

[1]   Bhuvaneshwari, S., Hettiarachchi, H. & Meegoda, J.N., Crop residue burning in India: Policy challenges and potential solutions. *Int. J. Environ. Res. Public Health*, **16**(5), p. 832, 2019.

[2]   Debabreta, B., (ed.), *Energy from Toxic Organic Waste for Heat and Power Generation*, Woodhead Publishing: Cambridge, UK, pp. 151–194, 2019.

[3]   Jha, P., Biomass characterisation and the application of biomass char for sorption of phenol from aqueous solutions. PhD thesis, IIT, Delhi, India, 1996.

[4]   Hamad, B., Noor, A.M. & Rahim, A.A., Removal of 4-Chloro-2-Methoxyphenol from aqueous solution by adsorption to oil palm shell activated carbon activated with K_2CO_3. *J. Phys. Sci.*, **22**(1), pp. 39–55, 2011.

[5]   Saeed, A.A.H. et al., Production and characterisation of rice husk biochar and kenaf biochar for value-added biochar replacement for potential materials adsorption. *Ecol. Eng. Environ. Technol.*, **22**(1), pp. 1–8, 2021.

[6]   Allaoui, S., Ziyat, M.N.B.H., Tijani, O.Q.N. & Ittobane, N., Kinetic study of the adsorption of polyphenols from olive mill wastewater onto natural clay: Ghassoul. *J. Chem.*, Article ID 7293189, 2020.

[7]   Siipola, V., Pflugmacher, S., Romar, H., Wendling L. & Koukkari, P., Low-cost biochar adsorbents for water purification including microplastics removal. *Appl. Sci.*, **10**(3), p. 788, 2020.

[8]   Nazal, M.K., Gijjapu, D. & Abuzaid, N., Study on adsorption performance of 2,4,6-trichlorophenol from aqueous solution onto biochar derived from macroalgae as an efficient adsorbent. *Sep. Sci. Technol.*, 2020. DOI: 10.1080/01496395.2020.1815778.

[9]   Dursun, G., Cicek, H. & Dursun, A.Y., Adsorption of aqueous phenol solution using carbonised beet pulp. *J. Hazard Mater.*, **125**, pp. 175–182, 2005.

[10] Hernadez, M.S., Tenango, M.P. & Mateos, R.G., (eds.), *Phenolic Compounds*, Intech Open: London, UK, pp. 420–443, 2017.

[11] Das, B., Characterisation of biomass/Agro residues and application of selected biomass for sorption of phenol from aqueous solutions. PhD thesis, SLIET, Longowal, India, 2016.

[12] Jha, P., Adsorptive findings on selected biomasses for removal of phenol from aqueous solutions. *Resource*, **8**(4), p. 180, 2019.

[13] Iyer, P.V.R., Rao, T.R. & Grover, P.D., *Biomass Thermo-Chemical Characterisation*, 3rd edn., Chemical Engineering Department, IIT Delhi: Delhi, India, p. 1, 2002.

[14] Nzihou, A., Thomas, S., Kalarikkal, N. & Jibin, K.P., (eds.), *Re-Use and Recycling of Materials: Solid Waste Management and Water Treatment*. Rivers Publishers: Lange Geer, Denmark, pp. 189–213, 2019.

[15] Trebal, R.E. (ed.), *Mass-Transfer Operations*, McGraw-Hill International Editions, pp. 565–612, 1980.

[16] Sarker, N. & Fakhruddin, A.N.M., Removal of phenol from aqueous solutions using rice-straw as an adsorbent. *Appl. Water Sci.*, **7**, pp. 1459–1465, 2017.

[17] Dass, B. & Jha, P., Batch adsorption of phenol by improved *Acacia nilotica* branches char: Equilibrium, kinetic and thermodynamic studies. *Int. J. ChemTech Res.*, **8**(12), pp. 269–279, 2015.

COATINGS AND SURFACE TREATMENT EFFECTS ON SOUND QUALITY IN CONCERT HALLS: A CASE STUDY OF THE BIBLIOTHECA ALEXANDRINA CONFERENCE HALL, EGYPT

SHAIMAA ABOU SHOUSHA, TAREK FARGHALY, IBRAHEM MAAROF & ONSY ABDELALEEM
Architectural Engineering Department, Faculty of Engineering, Alexandria University, Egypt

ABSTRACT
In concert halls, interaction between music and features of the hall happens continuously. As the concert hall is considered a box full of sound reflections and waves which get reflected all the time from and to every surface in the hall. However, most concert halls are not in the ideal state to achieve the optimum experience for the attendees in terms of sound quality. The most common reason is that they were designed for multiple different purposes such as conferences or speeches. This research aims to study the effect of material characterizations and surface treatment; such as porosity, absorption and reflection, and their effect on the sound quality in concert halls. This research consists of two parts: Part one is a review of concert hall design principles and the materials used in its furnishing and finishing, plus a review of every advantage and disadvantage of most used materials. The second part is a case study on a conference hall turned to a main concert hall in Bibliotheca Alexandrina, where sound measurements took, and software application was held after suggesting some changes to surface treatment and coatings in that hall. The results of the research have proven that changing surface treatment and coatings can help positively in turning any hall to a concert hall and enhancing it's the sound quality.
Keywords: materials characterization, surface treatment, coatings, concert halls, seats materials, wall coatings, sound absorption, sound reflection, experimental techniques, computer sound simulation.

1 INTRODUCTION
The acoustic evaluation of the classical concert hall has been an important issue for a long time. Mostly because the flow of the sound waves moving from the stage among audience seats can determine the sound quality in that hall.

Sound quality in the concert hall is largely affected by the architectural features, the interior design, finishing materials, and surface coatings.

Several studies have conducted an acoustic simulation analysis through computer simulation programs during the early planning phase. Barron and Kissner presented a possible acoustic design approach for multi-purpose auditoriums. Two acoustic models were simulated using computer simulation programs [1].

Materials have a very important effect on sound quality in concert hall acoustics. The same materials can be used in different spaces and make different effects if arranged in a different way or used in different coatings. This is because the orchestra needs all the reflection and reverberation that it can get to make the sound richer.

Some concert halls were not originally designed as a concert hall, and some of them are multipurpose halls were music concerts can be held but with very complicated and expensive sound systems [2].

The objective of this research is to enhance the sound quality in concert halls or in auditoriums converted to concert halls, by changing characterizations of materials used in furnishing the hall and surface coatings.

WIT Transactions on Engineering Sciences, Vol 133, © 2021 WIT Press
www.witpress.com, ISSN 1743-3533 (on-line)
doi:10.2495/MC210151

To achieve the objective of the research, it followed inductive and deductive methodology starting from the literature review to the results section. The research is divided into two sections, the first section is a theoretical framework using a literature review that discusses materials role in sound quality and discusses different material effects on reflecting and absorbing sound waves in concert halls. The second section is a two-phase section, where phase was a case study of the Bibliotheca Alexandrina conference hall where different sound measurements were taken live during rehearsals and concert, as an observation of the current situation, while in phase two a *full model simulation* was made by applying different material characterizations and different surface coatings and checking sound reflections and absorptions in the new state. The results of the simulation were then represented in the form of a table of new measurements; the research is then continued by a discussion of those results and measurements, and a general conclusion.

2 LITERATURE REVIEW

2.1 Sound quality principles in concert halls

Sound interior designs depend on the acoustic properties of the hall or the auditorium. The sound is mostly reflected by hard materials and absorbed by soft ones, gradually fading away as its energy is dissipated through coatings and different materials [2]. Reverberation or echo is particularly noticeable when a sudden bang continues to be audible for several seconds as in musical concerts [3].

Good acoustic design requires high understanding of the engineering characteristics of materials and the unique attributes of music. Then applying this knowledge to building construction to create a facility that allows listeners to have an outstanding aural experience [3].

Three design principles have been mentioned in concert hall design:

- Music should be optimally heard equally in every spot in the music hall (including an excellent "ring") throughout the entire performing venues, without having dead spots or echo spots [4].
- Sounds coming from outside the building needs to be minimized, if not entirely eliminated, to keep sound interference to a minimum. This includes the sounds of the traffic, or people on the streets, or the surrounding environment.
- If the building has different musical spaces, they shouldn't disturb or affect each other. Sounds created in one space should not disturb adjacent music spaces in the building [4].

A great concert hall has both sound isolation and good sound reverberation. These two qualities have to be taken into consideration while designing any music hall [5]. When a sound is produced in a concert venue, the reverberation of the hall can be timed until that sound has completely disappeared. Longer reverberation is desirable for a music space as opposed to a theatrical space, where clarity and intelligibility of speech is more important than sound quality [6]. Longer reverberation is difficult to achieve. To be successful, planners must create an adequate cubic volume and limit sound absorbing material. An additional consideration when building is to have adequate mass (weight) in all exposed surfaces; since it's important to have hard surfaces with no hollow spaces underneath [7].

2.2 Materials and coatings in concert halls

Designing good concert hall acoustics is a very delicate art as all elements unite to define the space in a very complex way. In this phase, I will explain the details behind the materials used in concert hall construction and coatings and how they affect each other acoustically; and which materials are recommended based on that [8].

Table 1:   Detailed comparison between hard acoustical materials used in finishes and furnishings and their properties and recommendations [9]–[11].

Material	Properties	Advantages	Disadvantages	Recommendations
Plaster	• Traditional material for ceilings and walls in concert halls. • Used in large areas and on surfaces.	• Used with extra thickness up to 1.5 to 2 inches. • Its stiffness and mass can resist vibration. • Low frequency absorption.	• Needs a very strong supporting structure. • Not very efficient in detailed decorative forms.	• Partly recommended • Better used in large spread areas than detailed areas. • Mostly used in simple ceilings and walls.
Gypsum board	• Gypsum board is a wallboard with a gypsum plaster core bonded to layers of paper fiber board.	• Used in detailed surfaces and ceilings. • Can be used in decorated ceilings in churches and auditoriums.	• Rarely used in concert halls as it has too much low frequency absorption. • Must be applied in two thick laminated layers in auditoriums.	• Partly recommended to be used in auditorium and concert halls with very large spaces and many frequencies.
Wood and wood panelling	• Wood panels applied on walls in concert halls and auditoriums for aesthetic reasons.	• Has an aesthetic appearance in concert halls. • Plywood can be bent with gypsum board over a curved wall or ceiling and be stressed so that it cannot vibrate strongly and will not exhibit so many low frequencies	• Thin wood panels with air behind it will allow unwanted low frequencies absorption due to panel vibration.	• Recommended to be used in concert halls. • Can't be used in thin panels and it should be backed with solid construction. • Should be used stressed to a back construction like gypsum board or plaster. • If used in curved system it should be rigidly supported.
Masonry	• Masonry materials such as brick, block, mason, and concrete. • Usually used in massive and thick way.	• Can fill large and huge spaces in massive and thick way.	• Masonry units porosity affects the sound reverberations significantly in the concert hall. • it absorbs middle and high frequencies but reflects low frequencies.	• Not recommended to be used in concert halls in large spaces and areas. • Its porosity should be coated and sealed before using it according to acoustical consultant recommendations.
Metals	• Metal sheets can be applied to walls and ceilings.	• Can be flexibly formed.	• Metal movement due to sound pulses can cause rattles. • It causes so many echoes and unwanted reverberations	• Not recommended to be used in concert halls but it can be limitedly used but fixed to solid core material.
Glass	• Glass pieces can be used in decorative purposes.	• Can be used in small decorative design and lightings.	• If used in large areas in concert hall walls or windows it can cause unwanted noise transmission.	• Not recommended to be used in large areas in concert halls. • Can be used in small decorative elements so they can provide some useful sound diffusion.

2.2.1 Hard Acoustical finishes and furnishings
A detailed comparison between hard acoustical materials used in finishes and furnishings and their properties and recommendations are shown in Table 1.

2.2.2 Soft acoustical finishes and furnishings

A detailed comparison between soft acoustical materials used in finishes and furnishings and their properties and recommendations are shown in Table 2.

Table 2: Detailed comparison between soft acoustical materials used in finishes and furnishings and their properties and recommendations [12]–[14].

Features	Properties	Recommendations
People	• People attending concerts. • Orchestra members. • Performers. • People are the most absorptive surface in concert hall.	• Room volume must be compared to the number of seats. • The most appropriate volume\seat ratio used in concert halls is 250:300 cubic feet per seat.
Seating	• Seats in concert halls are formed in rows and columns with different spacing between them. • It can absorb a generous amount of frequencies.	• Seating absorption should match the absorption of audience, so that when people are absent as in rehearsals or in low attendance, there will be a very limited change in reverberation times in the hall. • Seats should be upholstered on the cushion and the back rest. • The seat back shouldn't be upholstered. • Seat material should be a woven breathable fabric stretched over thick breathable foam padding.
Carpet	• Carpet used limitedly in concert halls in aisle runners.	• Carpet should rarely be used in concert halls. • It's not acoustically preferred. • It has a very strong absorption effect on frequencies. • It should be used only in aisle runners and in a thin layer with no padding.
Walls	• Back walls following the seats shape (curve), can reflect sound waves to the front, focused and strengthened, and with a time delay that makes strong echoes. • Balcony faces, can cause a serious echoes by doing the same thing unless it was treated. • Low wall at the cross aisle can do the same.	Two recommended approaches for wall treatments: • Making them very absorptive. • Shaping them to send sound reflections to the other walls. • According to the calculations of the reverberations number, it can be decided either absorption treatment is needed or it will add too much absorption to the room.
Draperies	• Draperies are used in concert halls, and they are generally absorptive.	• They should be rarely used in concert halls, especially around the orchestra platform, as they need the entire solid sound reflective backup.

3 CASE STUDY AND APPLICATION

3.1 Case study

The chosen case study is a conference hall in Bibliotheca Alexandrina, which was originally designed as a conference hall but by the time it became one of the most important concert halls in Alexandria.

The study is divided into two parts; the first part is studying materials and coatings used in the hall in the current situation. The second part is live sound measurement were taken during live concert and rehearsals.

3.1.1 Materials used and coatings

By studying the hall it was found that all seats are totally upholstered from all sides and contain thick sponge padding as shown in Fig. 1. Also a layer of thick carpets is used in furnishing the hall floor everywhere inside including aisles, runners, stairs, and the front of the hall as shown in Fig. 2. The ceiling is made mainly of Gibson Board panels as shown in Fig. 3. The stage is made of concrete, but after using the hall for musical concerts it was treated by a layer of rubber as shown in Fig. 4. Walls of the hall are coated with two materials. The upper part is coated by wood panels, while the lower part is coated by a wool layer as shown in Fig. 5. There are also very long and thick velvet draperies around the stage as shown in Fig. 6. The hall also contains some balconies.

3.1.2 Sound measurement

These are the sound measurements that were taken during a live concert and live rehearsals of a musical concert. It consists of instruments, solos, human voice, and choir using a sound level meter (SLM) as shown in Table 3.

Figure 1: Seats are totally upholstered from the back.

Figure 2: Thick carpets furnished all floors.

Figure 3: Hall ceiling and the bottom of the split level are totally covered with gypsum board panels.

Figure 4: Stage covered with rubber sheets.

Figure 5: Wall covered by a carpet layer.

Figure 6: Long and thick velvet draperies around the stage.

Table 3: Sound measurements were taken in 15 points in different spots.

Point no.	Tutti		Strings		Wind instruments		Opera soloist		Choir		Human voice talking	
	(Forte)	(Piano)	(Forte)	(Piano)	(Forte)	(Piano)	(Forte)	(Piano)	(Forte)	(Piano)	Forte	(Piano)
1	73	67	74	64	70	65	67	55	74	60	57	50
2	84	76	70	65	74	66	87	65	77	65	66	55
3	72	65	74	63	75	63	80	72	74	66	58	53
4	70	64	60	56	65	50	69	46	65	47	55	43
5	88	79	79	64	76	64	79	66	76	65	64	56
6	71	63	62	57	63	49	67	47	64	48	57	46
7	69	57	67	55	67	45	67	48	66	49	58	44
8	86	77	78	63	75	63	78	64	74	63	64	53
9	68	55	68	56	64	46	68	48	65	50	57	47
10	75	65	73	59	72	60	75	65	73	60	64	52
11	84	74	78	60	74	61	75	63	75	61	62	53
12	74	64	72	58	71	61	74	64	72	60	63	51
13	63	50	64	52	60	44	63	47	65	44	55	46
14	65	52	63	54	62	50	64	49	66	48	57	48
15	64	51	63	53	62	52	63	48	64	49	54	47
Concert	86	74	84	68	84	78	76	64	86	74	74	64

The measurements were taken by a sound level meter to measure the sound level reaching different points all over the hall as shown in Fig. 7. Points were chosen to cover every spot in the hall on the sides, front rows, back rows, and the upper part of seats in balconies.

4 DISCUSSION

After taking those measurements it was found that points no. 1,4,7 (Right side), and points no. 3,6,9 (left side) had a very low sound level specially in piano parts (low sound levels) as they are surrounded with carpet coatings on the lower half of the side walls and the back walls of the hall. And since carpets have a great absorption capacity, it causes high sound absorption around those points.

Points 13, 14, and 15, had the lowest sound level in the hall as they are at the far end of the hall and there is a huge split level elevated over all the audience steps in that area. So sound coming out of the stage travelling all the way through the audiences loses its power and level, then hits upper balcony walls and the huge ceiling before reaching those three points.

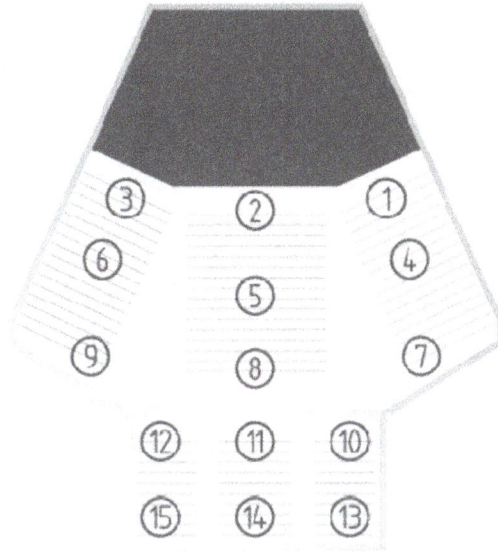

Figure 7: Points where measurements were taken live.

Points 2, 5, 8, and 11, had the best sound levels as they are at the centre of the hall and not surrounded by any extra absorption or reflection elements.

Points 10, 12 had an average sound level as they are partly surrounded by walls and carpets but have enough clearance between them and the stage allowing sound waves to reflect and reverb and reach the audience.

Fig. 8 shows the gradient of the sound level reaching every spot in the hall, where we can see that the red spots represent high sound level, and the grey and blue spots represent low sound level.

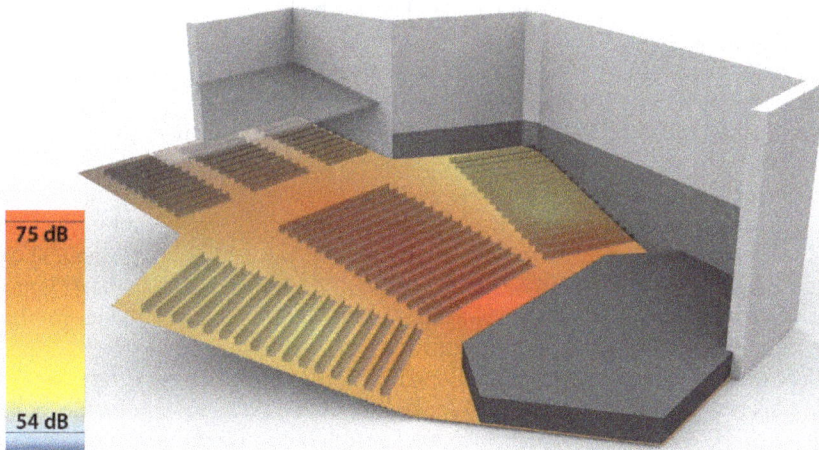

Figure 8:   Diagram produced by Rhino Grasshopper to represent weak and strong spots in the hall.

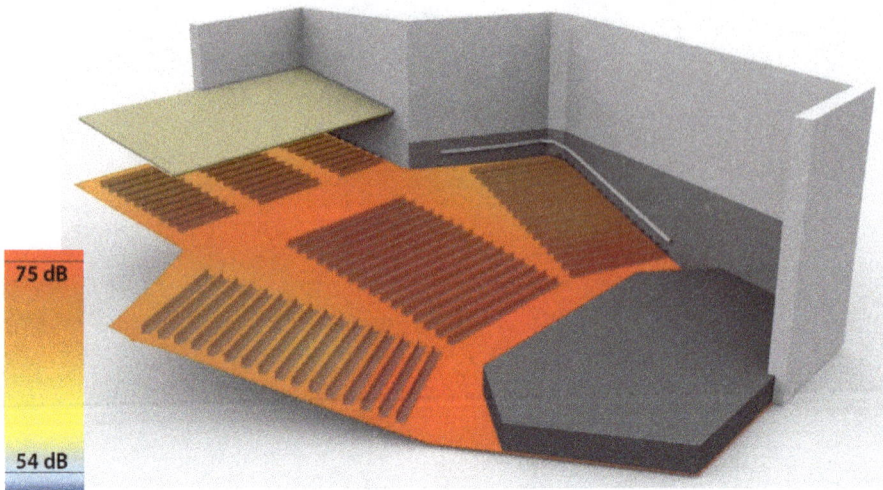

Figure 9:   Diagram produced by Rhino Grasshopper to represent progress in sound quality inside the hall after applying the metal part and the wooden cladding on the bottom of the split level.

4.1 Application

4.1.1 Materials recommendations

After studying the characteristics of the materials used in that case study, and studying other materials used in concert halls and their absorption and reflection effects; some of the material recommendations were concluded to solve current sound problems while using that conference hall as a concert hall [15].

a) Seats: The backs of all the seats are upholstered with a very thick layer of wool, and since wool is a very absorptive material which has high porosity [15], so I recommend replacing the back material with wooden backs which have relatively low porosity and high reflection effect.

b) Side walls: Side walls are divided into two halves, the upper half is covered by wooden panels, and the lower half is covered by a thick layer of carpets. This causes a very high absorption of sound coming from the stage reaching the audiences [16]. To solve that problem in the current situation, I suggested two solutions: The first one was adding wooden panels to the lower part of the wall, but that wouldn't be a practical solution for the existing situation as it would need a lot of major changes to the architectural interior and would incur high expenses. The second solution was adding some thin metal panels to the carpet layer where required in the shape of a handrail. We can determine its size and place by simulating sound level needed in that spot. I recommend that solution as it can be applied easily to the existing situation.

c) Balcony walls for the upper part of the hall: The split level covering the back area of seats has two problems. The walls of the balconies and the bottom of that level hanging over the back seats; which is made of large area of Gibson Board. As mentioned in Table 1 Gibson Board is not recommended to be used in concert halls in large areas and low heights as it is originally used in sound insulation purposes and has too much low frequency absorption. This leads to losing sound level power as appeared in last measurements as shown in Table 3. I recommend adding some wooden sheets strongly fixed to the ceiling to provide the required amount of sound reflection. A simple simulation was carried out to prove the result as shown in Fig. 8.

Fig. 9 shows the sound level in the hall after simulating adding suggested materials. We can obviously notice the difference in the sound quality reaching all the spots in the hall after the modification of the existing situation without major changes to the hall layout.

5 CONCLUSION

Materials and coatings have a major effect on acoustics in concert halls. The two factors in materials characterizations that strongly affect the sound quality are absorption and reflection; which depend mainly on the material porosity.

In concert halls to reach the perfect sound quality and level, there are some recommended materials to be used in coatings and surface treatment according to their absorption and reflection properties. Designers should take these into consideration during the design process. Many auditoriums or conference halls have been changed to be used as concert halls for musical events. This greatly affected the sound quality reaching the audiences in such mixed use facilities. There have not been previous studies that focused on the modification of auditoriums or conference halls to support a pleasing musical experience. In this research the aim is to prove that adding or removing some materials according to their characterizations of absorption or reflection can strongly affect sound quality in the whole venue; which is proved by simulation and measurements. This study uses a qualitative analysis using a methodological and applicable framework of the sound level variable, mainly by changing materials used in the most affected areas in the hall to improve sound quality.

ACKNOWLEDGEMENTS

Gratitude must be shown to God for helping in such a research. I would like to thank my supervisors Professor Tarek Farghaly, Professor Ibrahim Maarof, and Professor Onsy AbdelAleem for their dedicated support and guidance while working on this research. I also would like to thank my parents and my husband for supporting and encouraging me completing my work while facing many obstacles.

REFERENCES

[1]   Barron, M., Objective assessment of concert hall acoustics using temporal energy analysis. *Applied Acoustics*, **74**(7), pp. 936–944, 2013.
       DOI: 10.1016/j.apacoust.2013.01.006.
[2]   Samuel, J.S., Which material will most efficiently reflect sound? Project no. j0239, 2002.
[3]   Jeong, K., Hong, T., Kim, S.H., Kim, J. & Lee S., Acoustic design of a classical concert hall and evaluation of its acoustic performance: A case study, Republic of Korea. 22 May 2018.
[4]   Duplin, P.L., The sound principles behind concert hall acoustics. *That's Maths*. Jul 16, 2015.
[5]   Concert hall: How to get the perfect acoustics. (Part 2), Mar. 2016.
       https://soundzipper.com/blog/the-acoustics-of-a-concert-hall-part-2/.
[6]   Lokki, T. & Pätynen, J. Architectural features that make music bloom in concert halls, 22 May 2019.
[7]   Beranek, L.L., *Music, Acoustics and Architecture*, John Wiley & Sons: New York, US, 1962.

[8]   Field, D. & Ma, Z., Materials characterization. *An International Journal on Materials Structure and Behavior.* ISSN: 1044-5803, 2020.

[9]   Davenny, B., Materials and sound. 13 Jul. 2009.

[10]   Barron, M. & Dammerud, J.J., Stage acoustics in concert halls: Early investigations. Jan. 2006.

[11]   Gómez, J.Ó.G., Kahle, E. & Wulfrank, T., Shaping concert halls, "EuroRegio". 2016.

[12]   Barron, M., *Auditorium Acoustics and Architectural Design*, 1st edn. Taylor & Francis: London & New York, UK & US, 1993.

[13]   Principles of Acoustic Design, Music Center Goshen College. https://gcmusiccenter.org/about/acoustics/principles-acoustic-design/. Accessed on: 15 Apr. 2021.

[14]   Kawase, S., Factors influencing audience seat selection in a concert hall: A comparison between music majors and nonmusic majors. *Journal of Environmental Psychology*, **36**, pp. 305–315, 2013.

[15]   McCandless, D., Concert halls: Specifying for sound performance. Apr. 1990.

[16]   Arets, M. & Orlowski, R., Sound strength and reverberation time in small concert halls. *Applied Acoustics*, **70**, pp. 1099–1110, 2002. DOI: 10.1016/j.apacoust.

DESIGN AND SMART TEXTILE MATERIALS

AYMAN FATHY ASHOUR
College of Fine Arts and Design, University of Sharjah, UAE

ABSTRACT

This research discusses smart textiles and their implications in the design process. The aim is to equip designers with an understanding of the significant potential properties of smart textiles and inspire them to integrate these materials into the design process. The sensory, adaptive, responsive, and multifunctional qualities of smart textiles open a multidisciplinary scope of use that offers many alternatives in numerous industries, including construction, fashion, healthcare, sports, and automotive. Research and development for new and smart materials cross scientific boundaries, entering design territories and enhancing the quality of life and our environment. These textiles are created by integrating material properties with embedded technology to deliver aesthetically pleasing innovative functional qualities. However, the practices associated with the two disciplines of conventional textile design and advanced technology embedding often vary. Designers currently working with conventional textiles may be able to solve problems by utilizing smart materials and revisiting their design techniques to embrace the unique characteristics of the smart materials. The research findings suggest that designers should pursue appropriate approaches and contribute their productive skills and material expertise to utilize the smart textile design effectively.

Keywords: textile design, smart textiles, smart textile design.

1 INTRODUCTION

Materials that can sense and react to the surrounding environment have been labeled "smart." Advances in new and smart material design have reached new heights. Designers embed computational programmability in conventional materials like textiles to develop innovative material that is both functional and expressive [1], [2]. The influx of new approaches to designing with new and smart materials presents a wide range of opportunities [3]. However, the ways in which designers will utilize these materials have not been defined, as the possibilities for design expand in conjunction with the development of new and smart materials. How will designers embrace these new, smart materials and create designs that incorporate their unique qualities and support their interactions with humans and the environment? This question calls for a new understanding of these materials to facilitate their incorporation into innovative designs.

Smart textiles can react and adapt to stimuli sensed from the environment. Stimulus and response can involve various interactions, such as those based on magnetic, electrical, or thermal energy [4]. Smart textiles constitute a relatively new textile field that can produce new applications, such as wearable technology and technical textiles, by integrating design and technology [5]. Smart textiles are divided into two categories of smartness: passive and active. Passive smart textiles, viewed as the first generation of smart textiles, have functionality beyond the level offered by traditional textiles, while active smart textiles can adapt by changing their functionality due to a user action or in reaction to changes in the surrounding environment. These textiles may, for example, change shape or may store and regulate energy. While passive smart textiles typically depend on their structure to function, active smart textiles use some form of energy that can support sensors and actuators. The sensors and actuators allow the smart fabric to sense numerous stimuli from the surrounding environment and interpret and process this variety of inputs according to the surrounding environment [5]. The role of textile designers who have deep material

WIT Transactions on Engineering Sciences, Vol 133, © 2021 WIT Press
www.witpress.com, ISSN 1743-3533 (on-line)
doi:10.2495/MC210161

exploration experience and knowledge becomes vital to creating innovative forms of interactions between the designed object and its user [6].

2 LITERATURE REVIEW

Research work in the field of innovative smart textile applications is growing as scholars seek to discover the prospects presented by manipulating textiles down to the nanoscale for producing new and smart functionality, resulting in integrated intelligent systems that can wirelessly sense and communicate. Such systems offer a multifaceted set of design, material, and manufacturing challenges. This literature review highlights some of these challenges.

Textiles are characterized by being flexible and can be easily formed over objects. New textiles that are electrically conductive have been created for stretchy electronic structures that can be curved or shaped around curved shapes [7]. Stylios and Chen [8] introduced a smart textile product called Psychotextiles after studying the relationship between brain waves and design. Using electroencephalograms (EEGs), characteristics of patterns that impact specific emotions emerged, which were then designed into four pairs of smart, pattern-changing textiles for examination. They then developed a new thermochromic procedure to enable the construction of innovative yarns that can switch from one pattern into another when knitted into jacquard patterned textiles. This procedure was essential in achieving Psychotextiles [8].

Malm et al. [9] discussed conductive structures in which micron-size metal flakes of silver-coated copper were used as conductors in woven textiles, and the fabric's flexural stiffness was examined for performance, reliability, and durability. The authors discussed measuring and optimizing the electrical resistance and mechanical properties of different textile sensors. In another study, Stöhr et al. [10] tackled the concept of a thermal textile, which is based on a textile arrangement that shows temporal and spatial thermal contrast and can be applied in thermal communication applications.

In a publication detailing their research, Lis Arias et al. [11] discussed uncontrolled active molecules in the spraying of textiles and defined polymers that protect active components that permit the regulation of drug dosages, with encouraging results for clothes and in-home use, thereby developing bio-functional textiles. Vojtech et al. [12] addressed the problem of identifying a surface area that supports the improved functioning of electrically conductive polymer-based textiles. The researchers employed an electrochemical process to determine the resistance between two electrodes to compare surface areas of textile. ECG measurement and motion tracking were accomplished by developing a new hybrid soft textile electrode [12]. Furthermore, An and Stylios [13] reported that systematic measurements demonstrated that this hybrid textile electrode was capable of recording ECGs and motion signals synchronously, possibly providing a life-changing approach to constant health monitoring. Moradi et al. [14] suggested a solution of connecting e-textiles using embroidery; they researched the ways in which signal propagate control may be accomplished, enabling customized electromagnetic characteristics, such as filtering for wearable electronics. Moreover, energy generation was addressed through textile-based dye-sensitized solar cells in two studies – one by Juhász and Juhász Junger [15] and one by Juhász Junger et al. [16]. Juhász and Juhász Junger [15] were concerned with understanding the physical processes in the cell and optimizing them, while Juhász Junger et al. [16] introduced electro spun polyacrylonitrile nanofiber mats coated with a conductive polymer as collar cells.

With the fast growth of the Internet of things (IoT), which is affecting our lives, the way we work, our homes, and our communications, electromagnetic shielding is important, as it guards against emissions of electromagnetic frequencies as a health protective measure and to prevent possible hack attacks. Neruda and Vojtech [17] argued that smart textiles with their high flexibility and formability are well-suited to help address this issue.

3 A CALL FOR NEW MATERIAL UNDERSTANDING IN DESIGN

The designer's role calls for involvement in discovering new potentials for materials rather than only relying on known techniques to create product applications [18], [19]. Possible material designs are created through the collaborative interactions of individuals, processes, and the environment in general [20]. Smart materials have specific qualities that change and adapt over time and that can affect our experiences and the ways in which we interact with products [21]. Indeed, materials can be active and adaptive in various ways. In this context, adaptive is defined as material expressions in which distinctive qualities of computational materials adapt and react through interactions [22].

Adaptive as an approach for the expressive and performance characteristics of material-driven design offers exceptional prospects in design practice and research [23]. It enables an arena for designing emerging materials that involve the fields of biology, computation, and design [24]. How can designers utilize a textile, for instance, if they believe that it is adaptive to different situations in the surrounding environment? What are possible design solutions for situations in which an expected adaptive textile behavior may possibly evolve? As responses to such complex questions may not be forthright, changes in mindset and in approaches to dealing with smart materials towards more non-linear and open designs and use cases is needed.

3.1 Designing with smart textiles

In the case of dynamic textile surface patterns, the aesthetic manifestations and the functions can be joined together, unlike traditional static textile surface patterns, the aesthetic manifestation of which does not actively contribute to the function. The integration of information technology with textiles expands the possibilities for new use cases in which information supplies the building blocks for the surface pattern and vice versa, that is, the surface pattern functions as a textile interface that can, for example, display information [25]. Craftmanship in textile design is important to revive the interaction with the material to reinforce textile representations both functionally and aesthetically [26]. Materials are full of suggestions for their use if approached non-aggressively and the designer is receptive to new and innovative ideas. They provide a source of endless stimulation and inspire us in a most unexpected way [27], [28].

It is important to work directly with a material to be capable of developing it, as there is an important change in the articulation of textiles and how they may behave in conjunction with integrating information technology aspects and applications into the textile industry [29]. Some examples of experimental work with developing prototypes that merge new materials, textiles, and electronics are included in projects described in "E-broidery: Design and Fabrication of Textile-based Computing" [30]. For example, textiles have been used to display information, generate heat, and light up when dark [31], [32].

A textile surface pattern that reveals different aesthetic manifestations over time is called a dynamic textile surface pattern. An example is a surface pattern that appears to be striped initially then changes to appear checkered a moment later [33]. This demonstrates a type of textile surface pattern that needs to be designed differently than traditional textile surface patterns, which maintain the same state constantly, are designed [34].

The experience of designing a textile surface using materials that are capable of changing their appearance over time compared to designing a static textile surface pattern differs for the designer [35].

Apart from the design prospects, another challenge is the possible need to integrate electronic components. Consequently, textile designers and electronic engineers must collaborate on how to structure textiles that sense and react to different situations in the surrounding environment [36]. However, certain restrictions in the fields of textiles and electronics may conflict, e.g., the products of one are expected to be washable, while products from the other are negatively affected by water.

Moreover, challenges exist for producing electronic conductive yarn on an industrial level; arguably the most important is presented by the global economy, which principally encourages standardization, low prices, and production scale in industrial operations [37]. For example, when using hand looms, the weft builds the textile by turning the yarn at the edges and maintaining an uninterrupted construction of the woven structure. However, industrial weaving machines regularly cut the yarn at the edges, producing a textile that contains individual pieces of yarn that produce the textile structure, causing one cut length for each piece inlaid, which restricts the possible use of the traditional industrial weaving machines to integrate electronics at the weaving stage. Therefore, it is easier to work such smart materials in a hand loom upon starting the research for innovative smart textile materials [38].

Some techniques other than weaving are suitable for integrating electronics with textiles; still, significant limitations exist to the possible applications. A new approach to viewing textiles from an alternate perspective as a new modern craft will play an essential role in developing the potential of this new field of textiles. The limitations lie beyond the traditional questions of where and when a product will be made available and how long it will continue as a trend [39]. Time should be a crucial design variable in developing dynamic textile surface patterns. The time variable relates to spatial and temporal settings and can be characterized and developed considering the following factors: stimuli that initiate change, time, and duration of change, and finally, location where changes show. These considerations have to be considered when it comes to design methodology, materials, and techniques [40].

4 CONCLUSION

In an ever-evolving design arena, design research work that explores the understanding of new practices, such as how new potentials are detected during the production process in design practices that utilize smart materials, are critical [20]. The aim is to urge designers to work directly with smart textile materials during design development to extend their knowledge of various ways to design with smart textile materials and seek a better understanding of the great potentials of smart textiles. This article also discussed designing dynamic textile patterns [34].

In today's design environment, designers should work beyond the systematic traditional methods of the design process, in which the conceptualization of ideas comes first and then the concepts are translated into forms, functions, and materials represented subsequently in the design process [41], [42]. While new materials and technologies offer a wide range of potential innovate design solutions, designers need to pursue appropriate approaches and identify tools to work with new materials that are characterized by being changing or responsive to the surrounding environment.

Studying the advances in the smart textile research enhances our knowledge and allows us to identify relevant challenges, such as washability and user safety, two vital factors that need to be addressed. A change of culture is needed to support textiles with electrical embedded components. However, the textile supply chain is not ready to incorporate rapid change; moreover, designers and engineers need to collaborate. After all, we have an exhausted textile industry that is supplying products criticized for harming the environment, when intact prospects are available for the industry provided that smart textiles converse with traditional textiles to reach innovative applications that respond to contemporary needs.

Finally, the author calls for an experiential approach to designing smart materials and to developing the methods for designing with and for new materials, an experience that is fostered by smart materials that require new ways of thinking and designing.

REFERENCES

[1]   Vallgårda, A. & Redström, J., Computational composites. *Proceedings of the SIGCHI Conference on Human Factors in Computing Systems*, ACM, pp. 513–522, 2007.
[2]   Ishii, H., Lakatos, D., Bonanni, L. & Labrune, J.-B., Radical atoms: Beyond tangible bits, toward transformable materials. *Interactions*, **19**(1), pp. 38–51, 2012.
[3]   Studd, R., The textile design process. *Design Journal*, **5**(1), pp. 35–49, 2002. DOI: 10.2752/146069202790718567.
[4]   Van Langenhove, L. & Hertleer, C., Smart clothing: A new life. *International Journal of Clothing Science and Technology*, **16**(1–2), pp. 63–72, 2004. DOI: 10.1108/09556220410520360.
[5]   Koncar, V., *Smart Textiles and Their Applications*, Woodhead Publishing: Duxford, 2016.
[6]   Schneegass, S. & Amft, O., *Smart Textiles: Fundamentals, Design, and Interaction*, Springer: Cham, 2017.
[7]   Dils, C., Werft, L., Walter, H., Zwanzig, M., Krshiwoblozki, M. & Schneider-Ramelow, M., Investigation of the mechanical and electrical properties of elastic textile/polymer composites for stretchable electronics at quasi-static or cyclic mechanical loads. *Materials*, **12**, p. 3599, 2019.
[8]   Stylios, G.K. & Chen, M., The concept of psychotextiles: Interactions between changing patterns and the human visual brain by a novel composite SMART fabric. *Materials*, **13**, p. 725, 2020.
[9]   Malm, V., Seoane, F. & Nierstrasz, V., Characterisation of electrical and stiffness properties of conductive textile coatings with metal flake-shaped fillers. *Materials*, **12**, p. 3537, 2019.
[10]   Stöhr, A., Lindell, E., Guo, L. & Persson, N.-K., Thermal textile pixels: The characterisation of temporal and spatial thermal development. *Materials*, **12**, p. 3747, 2019.
[11]   Lis Arias, M.J. et al., Vehiculation of active principles as a way to create smart and biofunctional textiles. *Materials*, **11**, p. 2152, 2018.
[12]   Vojtech, L., Neruda, M., Reichl, T., Dusek, K. & De la Torre Megías, C., Surface area evaluation of electrically conductive polymer-based textiles. *Materials*, **11**, p. 1931, 2018.
[13]   An, X. & Stylios, G.K., A hybrid textile electrode for electrocardiogram (ECG) measurement and motion tracking. *Materials*, **11**, p. 1887, 2018.
[14]   Moradi, B., Fernández-García, R. & Gil, I., E-textile embroidered metamaterial transmission line for signal propagation control. *Materials*, **11**, p. 955, 2018.

[15] Juhász, L. & Juhász Junger, I., Spectral analysis and parameter identification of textile-based dye-sensitized solar cells. *Materials*, **11**, p. 1623, 2018.

[16] Juhász Junger, I. et al., Dye-sensitized solar cells with electrospun nanofiber mat-based counter electrodes. *Materials*, **11**, p. 1604, 2018.

[17] Neruda, M. & Vojtech, L., Electromagnetic shielding effectiveness of woven fabrics with high electrical conductivity: Complete derivation and verification of analytical model. *Materials*, **11**, p. 1657, 2018.

[18] Thota, H. & Munir, Z., *User-Centred Design. Palgrave Key Concepts: Key Concepts in Innovation*, Macmillan Publishers Ltd.: London, 2011.

[19] Barati, B., Karana, E. & Hekkert, P., Prototyping materials experience: Towards a shared understanding of underdeveloped smart material composites. *International Journal of Design*, **13**(2), pp. 21–38, 2019.

[20] Barati, B., Design touch matters: Bending and stretching the potentials of smart material composites (dissertation). Delft University of Technology: Delft, Netherlands, 2019.

[21] Giaccardi, E. & Karana, E., Foundations of materials experience: An approach for HCI. *Proceedings of the 33rd Conference on Human Factors in Computing Systems*, ACM: New York, pp. 2447–2456, 2015.

[22] Hobye, M. & Ranten, M., Behavioral complexity as a computational material strategy. *International Journal of Design*, **13**(2), pp. 39–53, 2019.

[23] Karana, E., Barati, B., Rognoli, V. & Zeeuw van der Laan, A., Material driven design (MDD): A method to design for material experiences. *International Journal of Design*, **9**(2), pp. 35–54, 2015.

[24] Bergström, J., Clark, B., Frigo, A., Mazé, R., Redström, J. & Vallgårda, A., Becoming materials: Material forms and forms of practice. *Digital Creativity*, **21**(3), pp. 155–172, 2010.

[25] Redström, M., Redström, J. & Maze, R. (eds), *IT+Textiles*, IT Press: Helsinki, 2005.

[26] Aktaş, B.M. & Mäkelä, M., Negotiation between the maker and material: Observations on material interactions in felting studio. *International Journal of Design*, **13**(2), pp. 57–67, 2019.

[27] Nimkulrat, N., Hands-on intellect: Integrating craft practice into design research. *International Journal of Design*, **6**(3), pp. 1–14, 2012.

[28] Adamson, G., *Thinking Through Craft*, Berg: Oxford, UK, 2007.

[29] Bai, Z., Innovative photonic textiles: The design, investigation and development of polymeric photonic fibre integrated textiles for interior furnishings (thesis). The Hong Kong Polytechnic University, 2015. http://hdl.handle.net/10397/35099.

[30] Post, E.R., Orth, M., Russo, P.R. & Gershenfeldt, N., E-broidery: Design and fabrication of textile-based computing. *IBM Systems Journal*, **39**(3–4), pp. 840–860, 2000.

[31] Worbin, L., Dynamic textile patterns, designing with smart textiles (thesis). Department of Computer Science and Engineering, Chalmers University of Technology and the Swedish School of Textiles, University of Borås, 2006.

[32] Dumitrescu, D. & Persson, A., Touching loops: Interactive tactility in textiles. *Proceedings of Futuro Textiel*, Kortrijk, Belgium, pp. 95–100, 13–15 Nov. 2008.

[33] Nimkulrat, N., Paperness: Expressive material in textile art from an artist's viewpoint (dissertation). Alto University: Helsinki, Finland, 2009.

[34] Worbin, L., Designing dynamic textile patterns (dissertation). The Swedish School of Textiles, University of Borås, 2010.

[35] Tan, J., Toomey, A. & Warburton, A., Craft tech: In hybrid frameworks for textile-based practice. *Journal of Textile Engineering & Fashion Technology*, **4**(2), pp. 165–169, 2018.

[36] Tan, J. & Toomey, A., *CraftTech: Hybrid Frameworks for Smart Photonic Materials*, Royal College of Art: London, UK, 2018.

[37] Tan, J., Kim, H. & Toomey, A., Sensory tactility: Designing interactive textiles for well-being. Presented at *11th Annual International Conference of Education, Research and Innovation*, Seville, Spain, 12–14 Nov. 2018.

[38] Tangsirinaruenart, O. & Stylios, G., A novel textile stitch-based strain sensor for wearable end users. *Materials*, **12**, p. 1469, 2019.

[39] Tan, J., POF smart textile design process. *Photonic Fabrics for Fashion and Interior. Handbook of Smart Textiles*, ed. X. Tao, Springer: Singapore, pp. 1005–1033, 2015.

[40] Petreca, B., Saito, C., Baurley, S., Atkinson, D., Yu, X. & Bianchi-Berthouze, N., Radically relational tools: A design framework to explore materials through embodied processes. *International Journal of Design*, **13**(2), pp. 7–20, 2019.

[41] Goodman-Deane, J., Langdon, P. & Clarkson, J., Key influences on the user-centred design process. *Journal of Engineering Design*, **21**(2–3), pp. 345–373, 2010. DOI: 10.1080/09544820903364912.

[42] Cross, N., *Engineering Design Methods: Strategies for Product Design*, 4th edn, John Wiley & Sons: Chichester, UK, 2008.

WASTEWATER MATTER: FROM ALGAE TO BIO-ALGAE PLASTIC 3D PRINTED FAÇADE ELEMENT

DEENA EL-MAHDY[1] & AHMED KHALED YOUSSEF[2]
[1]The British University, Egypt
[2]Cosign Group, Egypt

ABSTRACT

In Egypt during the extreme heat in summer, numerous amounts of air conditioners – that provide a cooler environment – are producing a huge amount of outlet wastewater. The continuous flow of this water causing great damage to buildings' façades. Therefore, the paper presents an innovative product solution made from algae that aim to reuse this wastewater as a self-watering landscape façade element that acts as an irrigation system. The prototype is designed from concept to manufacturing to implementation based on 3D printing with a bio-algae filament. With the dual algae ability in producing O_2 and absorbing CO_2, the fabrication follows a spiral engrave path to collect and cool the water droplet and ensure a smooth flow to be suitable for plantation. A path strategy is used during the printing for minimal structure supports aimed at saving unnecessary material waste and fabrication time. Solar radiation and water simulation are tested to measure the effect of the algae and to ensure the water fluidity from the AC tube till reaching the soil. The solar radiation results record a solar reduction from 316.43 to 80.71 kWh/m^2 after adding the algae panel to a building façade with a decrease of 6°C in the water temperature. The design demonstrates highly significant materials and resource savings, where no supports are needed during printing. The finding addresses the manufacturing of a low-cost algae product using cleaner technology as additive manufacturing. Given the alarming increase in the new industrial materials, algae will allow designers to explore their benefits regarding their O_2 production and CO_2 absorption, which will influence the façades to be smarter and sustainable using large-scale of PBR – photobioreactors – applications as a nature-based alternative to large glass surfaces with the potentials of additive manufacturing. This can reduce plastic production using fossil fuels to be eco-friendly.
Keywords: 3D printing, additive manufacturing, algae, bioprinting, digital fabrication, green façade, bioenergy.

1 INTRODUCTION

Since the last three decades, the extreme fluctuations in climate change have been rising the energy consumption rates globally at an escalating pace. The shortage in the resources caused by the rapid population growth also is increasing the demand for the use of industrial construction materials such as concrete and fired bricks causing non-comfort in the internal spaces. Since then, many countries have started to use recycled materials, cleaner resources energy, and sustainable process during manufacturing to ensure fulfilling the sustainable development goals for reaching sustainable cities. As a way to find new cleaner alternative and sustainable materials in construction, several studies integrated algae in architecture as a bioprinting material that can reduce carbon emissions and provide oxygen [1], [2]. For instance, with the importance of water management, many projects started to invest in providing more water to solve the shortage in some places. Thus, a new exploration of renewable and clean sources of water and energy started to gain prime importance. Consequently, a new concept of the regenerative city that depends on water, land, and energy in the respect of the environment grows widely, contributing to the resilience of the entire system [3]. The new agenda of Habitat III is to targeting cities to be highly energy efficient with low CO_2 emissions by increasing the relay on renewable energy sources through reusing and recycling the waste [4].

Climate change not only encompasses raising the average temperature, but it has a potential impact on many sectors in construction and water resources management [5]. Thus, the high changes in temperature did not affect only the indoor environment which increased the HVAC systems usage but the growth of the aquatic plants causing water pollution [6]. Therefore, the integration of green architecture became a good example to reduce carbon emissions and find renewable resources of energy to support the global climate action change summit. One of these green applications is biomaterials and algae which started to spread with the aid of technology [7]. Algae are one of the main components of aquatic organisms and plants that serve as the base of the food chain [8]. Both algae and microalgae can either live in colonies or grow with different patterns in freshwater, saltwater, wastewater, and rainwater [3], yet their nutrients are provided from the wastewater. Scientists and researchers have been widely studied and integrated algae in several fields up to now for plant physiology investigations because of their important role in oxygenation, filtration, and wastewater purification [5], [9]. In some countries, algae play a big role in the economy by the useful substances extracting from them. Microalgae also act as a useful biological indicator of environmental change and water quality monitoring [10], [11]. They also can be used in assessing ecological variations [12], addressing global environmental concerns including environmental degradation, energy demands [13], and producing light energy [1], [14].

In Egypt, one of the main noticed problems during summer is the numerous amounts of air conditioners (AC systems) that are working over the day producing a huge amount of outlet wastewater. The continuous water flow is causing great damage to the building façades besides the AC tubes that spoil the façade as shown in Fig. 1. As a way to prevent the damage of the façade caused by the water-waster, the paper presents an innovative sustainable product solution that aims to reuse the wastewater of the air conditioner as an automated watering platform for plants as a landscape element. The concept of the new device element stands for reusing algae as a local biopolymer material that is environmentally friendly and non-toxic with a low cost that adds more value to the design. Besides its availability in the Nile and its environmental benefits.

Figure 1: The damage of the wastewater from the AC tubes on the building façade in Egypt.

The paper documents the fabrication process of the 3D printing plant pot using bio-algae through our participation in (Algae lab workshop) that was held in Cairo under the supervision of Atelier Luma. The paper provides a method of using algae as a double skin in building façades for CO_2 filtration from the atmosphere. This element acts as an irrigation system that collects and cools the dropped water through an internal engraved spiral tube allowing water to flow around the outer skin till reaching the roots. The resulted prototype was tested on a building façade to measure its effect on solar radiation. A water simulation was done to verify the decrease in the water temperature from the air-conditioner tube until reaching the bottom of the pot. The reason for choosing the algae arises from two main factors which are; the high cost of the filament polymers and plastics that are neither locally produced nor widely available in the Middle East, and the non-sustainable filament that is hard to be scaled up. Using organic materials in 3D printing can change the production of plastic made from fossil fuels to be more environmentally friendly.

2 ALGAE AS A GROWN MATERIAL IN ARCHITECTURE

In Egypt, the River Nile, which is considered one of the main sources of water for irrigation, industrial and drinking purposes, contains different types of unused algae. Due to climate change and external pollution in this water, algae and biological life are affected [15]. Yet, the abounded availability of the organisms and macroalgal biomass in the marine ecosystem are considered as producers of clean source energy, oxygen, heat, and biological compounds biomass such as nutraceuticals, food, and fertilizers [7], [16], [17]. The global attention towards microalgae as a source of biofuel is mainly due to their ability to be converted into electricity and heat through the higher efficiency in the photosynthesis process [1]. This production process is activated by using sunlight and CO_2 to absorb carbon dioxide from the atmosphere transforming it into biomass and oxygen with the use of nutrients [2], [3].

With focusing on organic and ecological "grown material" studies and particularly on "algae-based" studies, the integration of algae in architecture has been performed at building and urban scale raising the threshold of resilience and built environment [3]. This integration opens new research dimensions to explore and reuse the renewable and waste materials that are merged in construction as a way to reduce the carbon emissions produced by the industrial materials. The application of living species as a construction material translated into algae-powered buildings is quite new in the world of architecture. Thus, this could serve as a multifunctional key to integrate smart cities with new eco-friendly technologies and concepts such as renewable energy production, waste valorization, circular economy, zero discharge, etc. Another usage of algal cells that they are being integrated into the design of solar thermal collectors because of their ability to improve indoor air conditioning systems, lighting systems, shading, etc. As an application, algae photobioreactors known as "PBR" became part of the building components that have been considered as a nature-based alternative to large glass surfaces [7]. One of the PBR benefits that they can provide shade during sunny days by cooling the building beside their long lifetime of 20 years, with an average of 1,480 kg of biomass production during this period [18]. The utilization of flat PBR panels on building façades can mimic the solar thermal unit in generating heat and efficiently absorb the UV light and other thermal light rays. Employing the production of bioenergy, preventing energy loss, recycling wastewater, and purifying air are among the major benefits.

Many applications of building façades used algae as an insulation and purification material to reduce environmental pollution, carbon emission, and produce oxygen.

Research into living architecture started to implement the biohybrid structure on the building façades [19] which is an integration of biological elements within structures to tackle the extreme environmental problems of our cities. The growing movement towards biohybrid structures potentially increases vegetation and green cover on building façades and rooftops. A bioprinting microalgae technique. Biomaterials as algae are represented as another group of naturally-derived biopolymers from marine biomass for 3D bioprinting for large-scale is developed for potential applications within architecture [19]–[21]. Thus, the appearance of additive manufacturing (AM) also known as 3D printing and rapid prototyping [22], [23] coupled with an interest in crafting traditions to develop biomaterials for architectural construction. One of the main AM advantages is the ability to reach homogenous hybrid structure components from mixed materials layer-by-layer [19] with no material waste. For instance, Ronald Rael with his group Emergent Object used non-conventional materials as sawdust, rubber, salt, and bioplastic as a 3D printing material [24].

From this perspective, the architects' role increased to bridge the gap between engineering and biotechnologist to develop new innovative products as a sustainable treatment. Eric Klarenbeek and Maartje Dros are Dutch designers who have developed a bioplastic made from algae as a solution to replace synthetic plastics over time. They cultivate algae then dry it to turn it into a filament material that can be used for 3D printers. After then the designers were invited to establish an open research and algae production lab at the Atelier Luma in Arles [25]. This lab developed an algae filament that can be used in 3D printers instead of PLA, which is used in our research.

Several applications of building façades and product design coupled algae with vegetation. In 2013, the BIQ (Bio Intelligent Quotient) house in Hamburg deployed Biohybrid large-scale structures on the façade including "PBR" Photobioreactors [19], which is considered the first bio-reactive green building façade known as solar-leaf. The PBR façades system is used as an adaptive shading screen and thermal insulation which in parallel utilizes the environmental advantages of algae to decrease carbon emissions. It also is regarded as a practicable solution to reduce the costs of this system [1], [3], [26] and to act as an acoustical element, economic, and aesthetics features. Through the treatment of wastewater, this façade used algal biomass and solar thermal heat to generate heat, biomass, biofuel, and bioenergy as renewable energy [2]. Bio reactive walls and PBR became a potential strategy for a ready scale-up of green walls [19] to transform CO_2 into biomass and oxygen through the photosynthesis process through sunlight and CO_2 [1], [3].

Another application that uses algae in 2019 is "H.O.R.T.U.S. XL Astaxanthing" which is considered the first 3D printed bio digital living sculpture and a new generation of biophilic architectural skin accessible on an urban scale. The project was a collaboration of many entities, where it was designed by ecologic-studio and fabricated by both CREATE group from the University of Southern Denmark and WASP Hub Denmark and developed by the Synthetic Landscape Lab at Innsbruck University. Inspired by coral morphology, a digital simulation was done to simulate the substratum growth. Based on large-scale 3D printers, the structure was made of triangular 3D printed units that were assembled to form hexagonal blocks. Cyanobacteria and micro-organisms were then injected into the printed units to absorb the carbon dioxide resulted from human breathing transforming it into oxygen and biomass through photosynthesis [27].

BANYAN eco wall project is one of the fully 3D printed green walls that comes with an embedded irrigation and drainage system designed by NOWLAB. The feature of this wall that it has rooted channels and a micro shower system that allows the feed of the plants with the precise amount of water and hydrates plants that live in and on the structure. The sophisticated drainage system of the prototype spans the whole structure [28]. The form of the wall was inspired by the organic components found in the plants such as roots, stems, and leaves. Without the AM technique, the complexity of the prototype form would be hardly achieved, injected the irrigation system inside, and controlled the size of the internal channels. The irrigation systems are fully controlled which ensures the amount of needed water in plants without human intervention need [29]. The project could be an inspiration for interior design or as a green vertical garden in façades and other forms of urban farming.

Apart from all the above-mentioned advantages of algae which need to be studied more deeply in practice, some challenges can be regarded applying PBR façade systems in buildings, which include a lack of sufficient regulation for construction, fire safety, and maintenance. Besides, the large-scale PBR needs complicated technology for biomass production, where regular maintenance is needed to ensure the cleanness of the living organism in the panel [1].

In Egypt, the massive availability of algae in the Nile and beaches is annoying, yet they are considered "biological waste" that is expensive to be disposed of. Tracing the gap where Egypt does not have any construction applications that use "algae" yet which is an abundant and renewable source, extracting "algae-based grown materials" can offer a new perspective for regenerative building construction to be integrated with the environment. Hence, the state of the art of our research is in using the wastewater from air conditioners in façades, unlike the previous algae applications. This is to highlight the importance of using this wastewater through a small-scale algal prototype which can be scaled up to be a façade element that functions for watering, filtration, purification, and aesthetic design as well.

3 MATERIAL AND METHOD

The paper highlights the importance of using the wastewater from air-conditioning for plantation which in parallel will reduce the carbon dioxide and act as an aesthetic element for more green façades. The methodology focuses on generating a prototype that is designed algorithmically using algae as filament material. The fabrication process of the prototype used a bioprinting technique that can change the plastic production in the world to be cleaner and environmentally friendly and provide more renewable resources. Fig. 2 shows the workflow of the fabrication that followed during the workshop which was two weeks long. The workflow started by sketching the prototype followed by a digital computational model on Rhino software for parameter controlling to prepare the model for printing. The digital model was converted to g-code layers to be ready for printing.

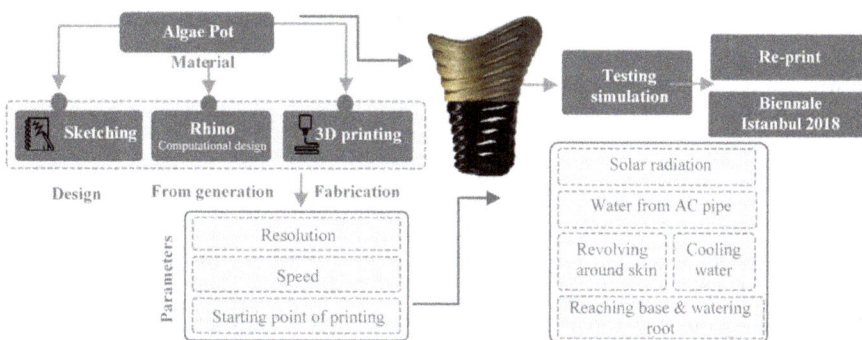

Figure 2: Method workflow process.

Different trials were done to adjust the speed of the nozzle and the temperature of the printer. The pot was then displayed at the design Biennial of Istanbul 2018, Arles 2018, Milano 2018, and Belgium 2019.

3.1 Material

The main material used is a ready-made algae filament -a bio-polymer mixture- whose ingredients were developed by Luma [30], confidential information under the umbrella of Algae lab a bio-laboratory in collaboration with Studio Klarenbeek and Dros. The production of this filament started in 2017 to explore the potential of growing both micro- and macro-algae by *Algae Lab*. The algae are mixed with biopolymers to produce a fully bio-sourced material that can replace non-biodegradable fossil oil-based plastics for a new circular production model through bio-fabrication as 3D printing [30]. The process of algae manufacturing was presented at the design Biennial of Istanbul as shown in Fig. 3.

Figure 3:   The filament algae process by Algae Lab during design Biennial of Istanbul 2018.

3.2 Design process and form generation

The pot has dimensions of 40×20 cm that its main function is to provide an automated process for watering the plants by collecting the wastewater and transfer it back to roots. The novelty of the spiral form followed several iterations on Rhino software to design a multifunctional piece that includes pours patterns and engraves from inside to ensure smooth water fluidity from the air conditioner tube until reaching the base. The spiral engraves act as an inner tube inside the pot which decreasing the outlet water temperature from the air-conditioner to be suitable for plantations While the patterns allow the airflow to pass to cool the water and act as an aesthetic form. The top of the pot is designed to be connected with the AC outlet pipe to guarantee to collect the water directly from the source as shown in Fig. 4. The pot diameter is larger in the top for providing a longer path to increase the water flow time and to ensure cooling the warm water from the tube before reaching the bottom.

3.3 Fabrication and machine

For the fabrication process, two models of 3D printers were used (Ultimaker 2) and (Monoprice MP Select Mini 3D Printer V2). The digital model of the prototype was divided into slices to be printed layer-by-layer each has its designed pattern through 3D slicer software – open-source – to be ready for printing. With a layer thickness of 0,1 mm,

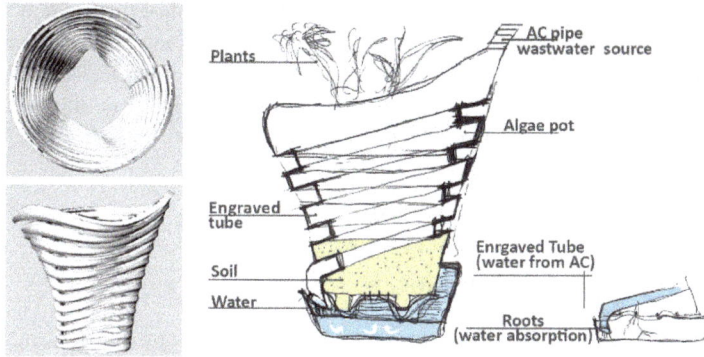

Figure 4:   The form of the pot shows the mechanism of the water flows till reaches the base.

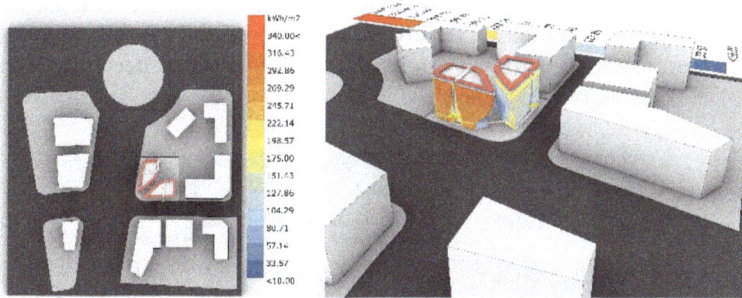

Figure 5: Radiation analysis of the building at the new administrative Capital in Cairo.

each layer was characterized by specific coordination (x, y, and z). Utilizing these data, the 3D printer traced each layer with a speed of 100 mm/s and a temperature of 220°C. The algae filament was not tested before in Egypt, which forces us to test the form first with its real scale using PLA filament to set any modification needed in the design before adding algae filament to the printer. A small-scale mock-up was first tested by using algae filament to examine the engraving in the form, temperature, speed, and accuracy. Several iterations were experienced to reduce the printing time and provide the material waste to print the pot with no support.

3.4 Mechanism and solar radiation test

A solar radiation test was run on the prototype by attaching it to a building façade to act as a double skin to run radiation analysis, thermal assessment, and shadow analysis to assess the design as a façade element. The test was done by using the ladybug tool in the Grasshopper plug-in in Rhino. This tool provides a direct calculation for the percentage of daylight exposure on a building façade through the year without any further calculations [31]. With a site selection at the New administrative capital in Cairo during July which is one of the highest degree summer months during the year, the shadow diagram helps in determining the best façade to place the PBR algae growing systems based on the number of hours exposed to the sunlight and the surrounding neighbouring which is our case western façade where the angle of the sun is lower than the southern façade as shown in Fig. 5. The direct sun exposure which is the main factor is to complete the close cycle of the PBR through photosynthesis. A water simulation was run to show the reduction of its temperature till reaching the end of the pot showing the circulation of the warm water cycle.

4 RESULTS AND DISCUSSIONS

4.1 Fabrication and printing mechanism

During the several iterations of setting up for parameters, speed, resolution, and direction of the printing, Fig. 6(a) shows the stability of the printing process using algae on the small-scale, where it started from the small diameter base till reaching upward the pot. It was observed that during the scaling up, some failures accrued because of the layers' weight where the materials needed more structural support. Accordingly, the pot was flipped upside down as shown in Fig. 6(b) to print the wide diameter of the top to ensure holding the weight without any extra structure support, which decreased the material waste. During the fabrication printing process, the form was tested with PLA before using the algae filament to ensure the consistency and smoothness of the water flow in the internal engraves. The complexity of the engraved form causes some errors during setting the speed.

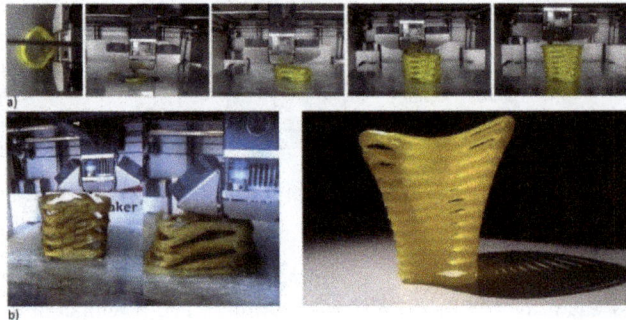

Figure 6:   (a) The printing process of the small-scale; and (b) The two printing iterations of flipping the pot.

4.2 Path planning strategy for minimal support

Path planning strategy was followed to achieve successful fabrication without any structure supports for saving unnecessary material waste and fabrication time. Thus, different iterations were tested for managing the minimal way of positioning the pot for printing as shown in Fig. 7(a). The printing started with the base as a starting point in the first iteration till reaching the top. It was observed that the deformation that occurred was a result of the form inclination where the radius of the base is smaller than the top. Hence, the layer was not hard enough to support and hold the additional weight of the materials upon each layer, which caused a pressure below causing collapse as shown in Fig. 7(b).

Consequently, in the second iteration, the direction of the pot printing was flipped where the top was printed first. This flip allowed the wider radius to act as a support during the printing to hold the weight of the above layers. This mechanism decreased the filament waste in the supports as shown in Fig. 7(c). For time reduction in the total nozzle movement length, a continuous path planning strategy was followed to decrease the distance traveled between subsequent space-filling for both the cantilever layer and the curves providing more time.

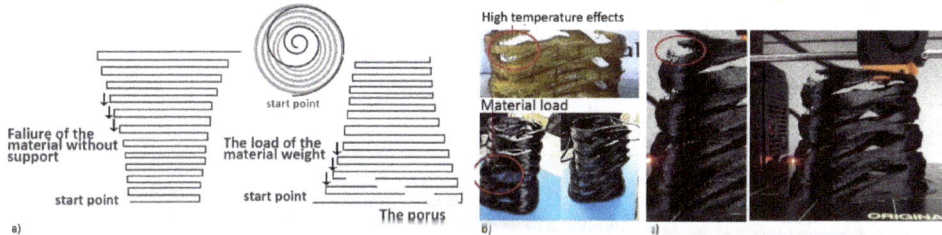

Figure 7:   (a) The spiral path; (b) The errors during printing; and (c) Minimum waste with no support.

Figure 8:   (a) The algae printed pot; and (b) Printing process during the Biennale exhibition, 2018.

4.3 Time reduction

Concerning the nozzle speed, resolution, and temperature, some failures occurred in the middle of the model while printing, where the speed started with 50 mm/s and an extrusion temperature of 280°C. It was observed that the higher temperature was, the higher the gaps in the materials that occurred. Therefore, the speed was increased to be 100 mm/s with a temperature of 220°C. Time reduction was one of the main targets to achieve the completed printing pot successfully with good quality. Yet, as known that 3D printing is a trial-and-error process where it needs more trials to get the final results. The estimation of the time was calculated before printing with a total of 18 hours when the speed was 50 mm/s. The prototype was first printed as a whole unit which was time-consuming because of the corruption that occurred during the printing. Therefore, the prototype was cut into two parts to be printed separately to decrease the failure, with a reduction in the resolution and increase in the speed which then decreases half of the time to be 9 hours in total for both parts. Fig. 8 shows the final assembly of the pot exhibited at the design Biennale in Istanbul in 2018, where the printer was set to print on-site to allow user interaction showing the whole printing process.

4.4 Solar radiation assessment and water temperature

Fig. 9 shows the comparison of the sun radiation before and after installing the algae PBR panel on the western building façade. The first model (a) shows a maximum radiation scale of 316.43 kWh/m^2. Taking into consideration that the optimum thermal radiation is between 80 to 100 kWh/m^2, the radiation in the model (b) after adding the algae panel was decreased to reach lower radiation that varies between 80.71 and 104.29 kWh/m^2. This will allow using less AC which in parallel will decrease the energy cost and the consumes excess energy.

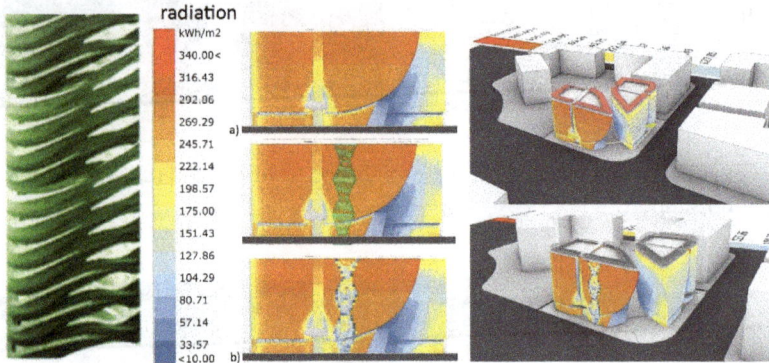

Figure 9:   The sun radiation on the western façade before and after adding the algae panel.

Fig. 10 shows the fluidity of the water through the spiral engraves that allow water to flow around the skin and cool the water droplets till reaching the roots. The length of the spiral path (60 cm) with the wide diameter at the top allowed more travel time to the water which decreased its temperature around 5–6°C from 42°C to be 36°C which means (1 degree each 10 cm. The small opening patterns in both the model and the panel prevent the solar radiation to pass inside, yet it still completes the cycle of photosynthesis to filter the carbon dioxide.

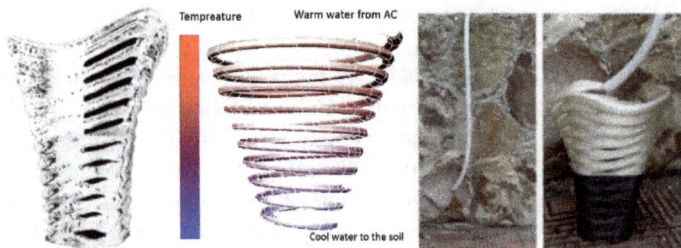

Figure 10: The decrease in the water temperature through the engraved spiral inner tube.

5 CONCLUSION

The paper highlights the importance of algae as a grown material showing several architectural applications integrated into building façades. The benefits of algae can be seen as producers of clean source energy, oxygen, heat, and biomass and also for CO_2 absorption through the photosynthesis process. Potentially, the paper demonstrated the possibility of a 3D printing algae device for using wastewater and attempted to generate an automated watering system on a small-scale pot that can then be applied up-scaled as bio-reactive green façades for self-controlled growth on buildings and urban scale.

The solution in this paper mainly targets rich places with algae to reuse this biomaterial in 3D printed construction elements. The added value of the new prototype lies in its automation system, affordability, property, flexibility, and sustainability. The spiral design played the main role in cooling the wastewater from the AC tube till reaching the soil. The fabrication process elaborated the direction of the printing which used a minimal amount of material without any structural support which provided unnecessary waste materials. The paper reached a process that could be followed regarding our different usage of algae. Given the alarming increase in the current substitution of traditional materials with new industrial materials, we believe that the findings will impact the construction building façades and will open new doors for more innovative large-scale PBR applications. This research is considered a first step to integrating algae in architecture where further research will be followed to study the LCA and thermal tests. Further research is also needed to monitor the effects of PBR-based microalgae façade design on its performance. On an urban scale, PBR can be inserted directly into the existing buildings, performing the dual function of absorbing CO_2 and producing biomass where needed most. Therefore, we assumed that in the future, these self-watering biohybrid systems could be custom fabricated with large scales using AM techniques from biomaterials. Further environmental and life cycle assessment of the algae is needed for future research. The CO_2 efficiency needs to be measured in future studies. A cost and benefit analysis to compare the tangible and intangible benefits of PBR system is needed. Finally, the future of 3D bioprinting towards bio digital materials will allow anyone to customize and 3D print any element using any type of materials as simple as baking bread.

ACKNOWLEDGEMENTS

The authors would like to thank Atelier LUMA ARLES for affording the algae filament during the workshop that was held in Cairo and hosted by Creative Hub and Creative Mediterranean at Art Jameel co-working space. Special thanks to the facilitators: Giulio Vinacci (the design expert) and Henriette Waal (Research Director of Atelier Luma) who displayed the results in several exhibitions all over the world.

REFERENCES

[1]   Talaei, M., Mahdavinejad, M. & Azari, R., Thermal and energy performance of algae bioreactive façades: A review. *J. Build. Eng.* **28**(Jan. 2019), 2020. DOI: 10.1016/j.jobe.2019.101011.

[2]   Biloria, N. & Thakkar, Y., Integrating algae building technology in the built environment: A cost and benefit perspective. *Front. Archit. Res.*, 2020. DOI: 10.1016/j.foar.2019.12.004.

[3]   Violano, A. & Cannaviello, M., Green-algae resilient architecture. *Rivista, Sustain. Mediterr. Constr.*, (Feb). pp. 142–149, 2020. https://doi.org/http://hdl.handle.net/11591/414061.

[4]   McDonough, W. & Braungart, M., Cradle to cradle: Remaking the way we make things. *Farrar, Straus and Giroux*, 2010. ISBN: 0865475873.

[5]   Haroon, A.M. & Hussian, A.M., Ecological assessment of the macrophytes and phytoplankton in El-Rayah Al-Behery, river Nile, Egypt. *Egypt. J. Aquat. Res.*, **43**(3), pp. 195–203, 2017. DOI: 10.1016/j.ejar.2017.08.002.

[6]   Short, F.T., Kosten, S., Morgan, P.A., Malone, S. & Moore, G.E., Impacts of climate change on submerged and emergent wetland plants. *Aquat. Bot.*, **135**, pp. 3–17, 2016. DOI: 10.1016/j.aquabot.2016.06.00.

[7]   Talebi, A.F., Tabatabaei, M., Aghbashlo, M., Movahed, S., Hajjari, M. & Golabchi, M., Algae-powered buildings: A strategy to mitigate climate change and move toward circular economy. *Smart Village Technology: Concepts and Developments*, eds S. Patnaik, S. Sen & M.S. Mahmoud, Springer International Publishing: Cham, 2020, pp. 353–365. DOI: 10.1007/978-3-030-37794-6_18.

[8]   Napiórkowska-Krzebietke, A., Hussian, A.-E.M., El-Monem, A.M.A. & El-Far, A.M., The relationship between phytoplankton and fish in nutrient-rich shallow lake qarun, Egypt. *Oceanol. Hydrobiol. Stud.*, **45**(4), 2015. DOI: 10.1515/ohs-2016-0045.

[9]   Km, K., Eshag, A., Sm, E., Sa, S. & Oa, A., Biodiversity & endangered species seasonal variation of algae types, counts and their effect on purified water quality case study: Al-Mogran and burri plants, Khartoum state. *Biodivers. Endanger. Species*, **2**(2), pp. 2–5, 2014. DOI: 10.4172/2332-2543.1000122.

[10] El-Naghy, M., El-Shahed, A., Fathy, A.A. & Badri, S.A., Water quality of the river Nile at minia, Egypt as evaluated using algae as bioindicators. *Egyptain J. Phycol.*, **7**(2), pp. 141–162, 2006. DOI: 10.21608/egyjs.2006.114160.

[11] Çelekli, A. & Külköylüoğlu, O., On the relationship between ecology and phytoplankton composition in a karstic spring (Çepni, bolu). *Ecol. Indic.*, **7**(2), pp. 497–503, 2007. DOI: 10.1016/j.ecolind.2006.02.006.

[12] Hamed, A., Biodiversity and distribution of blue-green algae/cyanobacteria and diatoms in some of the Egyptian water habitats in relation to conductivity. *Biology*, 2008. https://doi.org/ID: 55266079.

[13] Jacquin, A.-G., Brulé-Josso, S., Cornish, M.L., Critchley, A.T. & Gardet, P., Selected comments on the role of algae in sustainability. *Sea Plants, Advances in Botanical Research*, vol. 71, ed. N.B.T.-A. in B.R. Bourgougnon, Academic Press, pp. 1–30, 2014. DOI: 10.1016/B978-0-12-408062-1.00001-9.

[14] Wong, Y.S. & Tam, N.F.Y., *Wastewater Treatment with an Algae*, Springer: Berlin, 1998. DOI: 10.1016/j.biortech.2016.09.106.

[15] El-din, S.M.B., Hamed, A.H.S., Ibrahim, A.N., Shatta, A.M. & Salah, A., Phytoplankton in irrigation and draining water canals of east Nile delta of Egypt. *Glob. J. Biol. Agric. Heal. Sci.*, **4**(2), pp. 56–60, 2015. ISSN: 2319-5584.

[16] Jaiswal Kumar, K. et al., Hydropyrolysis of freshwater macroalgal bloom for bio-oil and biochar production: Kinetics and isotherm for removal of multiple heavy metals. *Environ. Technol. Innov.*, **22**, pp. 1–14, 2021. DOI: 10.1016/j.eti.2021.101440.

[17] Veluchamy, C. & Palaniswamy, R., A review on marine algae and its applications. *Asian J. Pharm. Clin. Res.*, **13**(3), pp. 21–27, 2020. DOI: 10.22159/ajpcr.2020.v13i3.36130.

[18] Mata, M.T. et al., Carbon footprint of microalgae production in photobioreactor. *Energy Procedia*, **153**, pp. 426–431, 2018. DOI: 10.1016/j.egypro.2018.10.039.

[19] Malik, S. et al., Robotic extrusion of algae-Laden hydrogels for large-scale applications. *Glob. Challenges*, **4**(1), pp. 1–12, 2020. DOI: 10.1002/gch2.201900064.

[20] Liu, J., Sun, L., Xu, W., Wang, Q., Yu, S. & Sun, J., Current advances and future perspectives of 3D printing natural-derived biopolymers. *Carbohydr. Polym.*, **207**, pp. 297–316, 2019. DOI: 10.1016/j.carbpol.2018.11.077.

[21] Axpe, E. & Oyen, M.L., Applications of alginate-based bioinks in 3D bioprinting. *Int. J. Mol. Sci.*, **17**(12), pp. 1–11, 2016. DOI: 10.3390/ijms17121976.

[22] Jiang, J., Path planning strategies to optimize accuracy, quality, build time and material use in additive manufacturing: A review. *Micromachines*, **11**(633), pp. 1–20, 2020. DOI: 10.3390/mi11070633.

[23] Jiang, J., A novel fabrication strategy for additive manufacturing processes. *J. Clean. Prod.*, **272**, pp. 122–916, 2020. DOI: 10.1016/j.jclepro.2020.122916.

[24] Rael, R. and Fratello, V.S., *Printing Architecture: Innovative Recipes for 3D Printing*, Princeton Architectural Press, 2018. ISBN: 1616896965.

[25] Morris, A., Dutch designers convert algae into bioplastic for 3D printing. *Dezeen*, 2017. https://www.dezeen.com/2017/12/04/dutch-designers-eric-klarenbeek-maartje-dros-convert-algae-biopolymer-3d-printing-good-design-bad-world/. Accessed on: 19 Apr. 2020.

[26] Wang, Y. et al., Perspectives on the feasibility of using microalgae for industrial wastewater treatment. *Bioresour. Technol.*, **222**, pp. 485–497, 2016. DOI: 10.1016/j.biortech.2016.09.106.

[27] Villa, D., The world's first 3D printed biodigital living sculptures featuring at the centre Pompidou in Paris. *WASP*, 2019. https://www.3dwasp.com/en/the-worlds-first-3d-printed-biodigital-living-sculputers-featuring-at-the-centre-pompidou-in-paris/. Accessed on: 11 Jun. 2020.

[28] BigRep, BANYAN eco wall – the world's first fully 3d printed, irrigated green wall. *BigRep*, 2019. https://bigrep.com/posts/banyan-eco-wall/. Accessed on: 19 Apr. 2020.

[29] Boissonneault, T., NOWLAB's 3D printed BANYAN ECO WALL integrates self-watering plant feature (3D), Printing Media Network, 2019. https://www.3dprintingmedia.network/nowlab-3d-printed-banyan-eco-wall/. Accessed on: 19 Apr. 2020.

[30] Luma, A., Can we use design to bridge the worlds of culture, science and the industry? *Algae Platform*, 2019. https://atelier-luma.org/en/projects/algae-platform. Accessed on: 26 Dec. 2020.

[31] Maksoud, A., Adel, M., Majdi, A. & El-mahdy, D., Generating optimum form for vertical farms skyscrapers in UAE. *Int. J. Eng. Res. Technol.*, **10**(3), pp. 689–700, 2021. ISSN: 2278-0181.

TIN(II)-CONTAINING FLUORIDE ION CONDUCTORS: HOW TIN MULTIPLIES THE FLUORIDE ION CONDUCTION BY UP TO THREE ORDERS OF MAGNITUDE

GEORGES DÉNÈS, ABDUALHAFED MUNTASAR, M. CECILIA MADAMBA & JUANITA M. PARRIS
Laboratory of Solid State Chemistry and Mössbauer Spectroscopy,
Department of Chemistry and Biochemistry, Concordia University, Canada

ABSTRACT

Electrical conduction by the motion of ions in the solid state was considered to be impossible for a long time. When Michael Faraday discovered that solid lead(II) fluoride conducts electricity when heated, it was an anomaly and the conduction mechanism remained ununderstood for long. Nowadays, several ions of small size and low charge, such as H^+, Li^+ and F^-, are known to be highly mobile in solid compounds provided the crystal structure contains pathways that make possible easy ionic motion. We have prepared new compounds, $PbSnF_4$ and $BaSnF_4$ and other phases in the PbF_2-SnF_2 system that increase the conduction efficiency by up to three orders of magnitude relative to the corresponding MF_2. Solid state fluoride ion batteries (FIBs) promise high specific energy and thermal stability. High potential solid state rechargeable fluoride ion batteries are being designed and $BaSnF_4$ has been used as one of the ingredients to form an interlayer solid electrolyte with high conductance. In the current work, we have analyzed the role of divalent tin in these structures, in terms of crystal structure and of local bonding to fluorine, in an effort to understand how it can result in such a high enhancement of the conductivity. Diffraction methods were used to investigate the crystal structures and ^{119}Sn Mössbauer spectroscopy for studying the mode of Sn-F bonding and the electron configuration of tin.
Keywords: ionic conductivity, fluoride ion mobility, divalent tin, lone pair stereoactivity, X-ray diffraction, Mössbauer spectroscopy.

1 INTRODUCTION

Electrical conductivity in solids was for long thought to be a property of metals only, in contrast with ionic compounds that were known to be insulators. When Faraday discovered in 1838 that lead(II) fluoride PbF_2 becomes an electrical conductor at high temperature, the reasons for this unusual behavior were not understood at the time [1]. Ions were not expected to be mobile in solids, where they are held in deep potential wells in the ionic lattice. Later studies lead to the conclusion that ionic compounds conduct electricity by the motion of ions in the liquid state and in solutions, where ions are free to move. Many more recent studies of the conductivity of PbF_2 have shown that Faraday's observations were correct and that the conductivity of PbF_2 increases substantially at the $\alpha \rightarrow \beta$ (orthorhombic → cubic) phase transition [2]. The cubic phase has the fluorite-type crystal structure and the high conductivity is attributed to the high mobility of the fluoride ions by means of Frenkel defects, and at higher temperature to the presence of a diffuse superionic transition resulting in a "sublattice melting" [3]. However, the mobility, of the fluoride ions is related to the room available to lodge temporarily the fluoride ions in the Frenkel defect model. One of us (GD) has shown that, if only the size of the interstitial sites were the main factor for determining the ability of the fluoride ions to move, the conductivity of BaF_2 would be much higher than that of β-PbF_2, and it is the reverse that is observed [4]. It was also shown by one of us (GD) that combining the fluorite-type BaF_2 with stannous fluoride SnF_2 results in $BaSnF_4$ with a substantial increase of the conductivity, by three orders of magnitude [5]. The SnF_2-based fluoride ion electrolytes $MSnF_4$ (M = Ba, Pb) have now reached the stage where they are being considered for application in room-temperature solid-state fluoride ion batteries [6], [7].

The aim of the present work is (i) to study how SnF_2 manages to increase the conductivity of the fluorite type MF_2 fluorides by such a dramatic amount (a thousand times), and (ii) how to design a method for determining rapidly whether a tin(II)-containing compound may be an electronic conductor or an ionic conductor.

2 THEORETICAL BACKGROUND REGARDING MÖSSBAUER SPECTROSCOPY RELATED TO ELECTRONIC STRUCTURE AND BONDING IN DIVALENT TIN

Crystallography uses X-ray or neutron diffraction to study of the whole crystal lattice. On the other hand, Mössbauer spectroscopy is a local probe. It is a nuclear spectroscopy, and therefore, it probes only specific nuclides, such as [119]Sn in this work. Therefore, by use of [119]Sn Mössbauer spectroscopy, we can study tin in compounds, and the way it interacts with its neighbors and with the entire solid lattice. The nuclear spins on [119]Sn are 1/2 in the ground state and 3/2 in the first excited state. The ground state has no quadrupole moment, therefore it remains unsplit ($|\pm 1/2\rangle$) in the absence of a magnetic field, even if an electric field gradient (e.f.g.) is present, while the first excited state has a quadrupole moment, that interacts with an e.f.g. to give rise to two sublevels, $|\pm 1/2\rangle$ and $|\pm 3/2\rangle$. Therefore, the resonant absorption of γ-rays by the [119]Sn nuclide for each tin site will give either a doublet ($|\pm 1/2\rangle \rightarrow |\pm 1/2\rangle$ and $|\pm 1/2\rangle \rightarrow |\pm 3/2\rangle$ transitions), or a singlet ($|\pm 1/2\rangle \rightarrow |\pm 1/2\rangle$ only), depending on whether there is an e.f.g. acting at the nucleus or not. No six line hyperfine magnetic field was observed since tin is diamagnetic in all its oxidation states, there was no transferred field in any of the compounds studied since they are diamagnetic and no external magnetic field was applied. The line position is called *isomer shift δ*, and it is a function of the amount of valence *s* electron density acting at the nucleus. This makes it a function of the oxidation state of tin, of the mode of bonding and of the electronegativity of the elements bonded to tin. To a lesser extent, it is also a function of temperature (*second order Doppler shift*). In Fig. 1(a), the Sn^{2+} stannous ion gives a single peak at high isomer shift (ca. 4 mm/s). The line is a singlet because the lone pair is located on the native *5s* orbital that is spherical and therefore it generates no e.f.g. The fact that the lone pair is purely *5s* makes it that its electron density is unshared and therefore there is a non-negligible amount of *5s* electron density acting at the nucleus, resulting in a high isomer shift. An e.f.g. due to lattice distortion is possible, however its effect on the spectrum is much weaker. Fig. 1(b) shows the case of tin(II) covalently bonded to fluorine. The stereoactive lone pair (non-bonded electron pair) at tin generates a very large e.f.g., hence a large doublet is observed. Since the *5s* electron density is shared between the lone pair and the bonds, it is further away from the nucleus than in the case of ionic bonding, and this generates a smaller isomer shift. Fig. 1(c) is the spectrum of $CaSnO_3$. It contains Sn^{4+} ions that have completely lost their valence shell, therefore there is no more *5s* electron density acting at the nucleus. It results in a much smaller isomer shift, that is taken as reference of isomer shifts, hence 0 mm/s, since all isomer shifts are referenced to $CaSnO_3$ at ambient temperature. It gives a single line since $CaSnO_3$. has the perovskite structure where the Sn^{4+} ions are in an octahedral site. It results from the above that the information obtained for each tin site are (i) its oxidation number, and in this study, for tin(II), (ii) the identification of the non-metal bonded to tin, (iii) the tin site distortion, and (iv) the tin type of bonding (ionic or covalent). Fig. 1 shows the influence of oxidation number and type of bonding on the Mössbauer spectrum. In addition, the lattice strength can also be estimated. Resonant

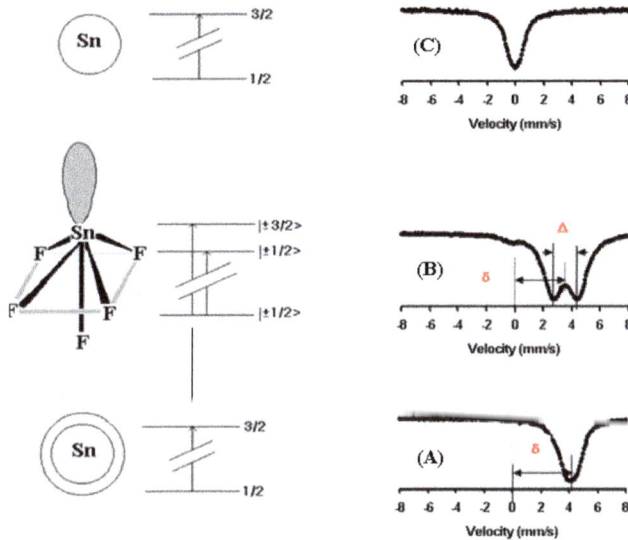

Figure 1:   Mössbauer spectra for (a) Ionic Sn^{2+}; (b) Covalently bonded Sn(II) in $BaSnF_4$; and (c) $CaSnO_3$.

absorption of γ-photons is necessary in order to produce a Mössbauer spectrum, and less and less of this occurs when temperature increases and when the lattice is weak, due to phonons. Therefore, a weak lattice will result in a weaker spectrum.

3 MATERIAL PREPARATION AND CHARACTERIZATION

The following starting materials were used: SnF_2 99% from Ozark Mahoning, $BaCl_2 \cdot 2H_2O$ analytical grade from American Chemicals, BaF_2 99% from Allied Chemicals, PbF_2 99.9% from Alfa and Dye Corporations, HF 40% aqueous solution from Mallinckrodt, and doubly distilled or deionized water. All crystalline reactants were checked by X-ray powder diffraction and were found to have only the expected peaks. In addition, SnF_2 was checked by DTA and identified by the α → γ transition at 150°C–160°C and melting point at 215°C [8]. The degree of hydration of $BaCl_2 \cdot 2H_2O$ was checked by TGA and found to be 2.04. The production of anhydrous $BaCl_2$ was verified by dehydrating the dihydrate at 140°C, significantly above the reported dehydration temperature of 113°C to ensure complete dehydration [9]. Total loss of water was confirmed by the mass loss of 14.71% (theoretical for $2H_2O$ = 14.75%).

All the materials investigated in this study were prepared either by precipitation or by solid state reactions in dry conditions. Solid state reactions were carried out by heating intimately mixed powders of the reactants in a sealed copper tube under dry nitrogen, according to the method developed by one of us (GD) [10]. The $MSnF_4$ compounds were prepared by heating stoichiometric amounts of SnF_2 and of MF_2 at 250°C for α-$PbSnF_4$ and at 500°C for $BaSnF_4$ according to the conditions determined earlier by one of us (GD) [11]. The α-phase of $PbSnF_4$ was also prepared by precipitation upon adding a solution of $Pb(NO_3)_2$ to a solution of SnF_2 upon stirring, for a molar ratio very rich in SnF_2 ($SnF_2/Pb(NO_3)_2$ = 4:1) [11].

X-ray powder diffraction was carried out by use of a Philips PW1050 diffractometer that had been automated with the Sie112 Sietronics® system from Difftech. This allowed a phase identification of phases already known, by comparison with the diffraction patterns of starting materials and other possible side products already collected in our laboratory and by use of the μPDSM Micro Powder Diffraction Search Match® from Fein-Marquat. Only the phases of interest in this study were subjected to further analysis.

The Mössbauer spectra were recorded using the following set-up. The source was a nominally 25 mCi $Ca^{119m}SnO_3$ γ-ray source from Ritverc GmbH. Isomer shifts were referenced relative to a standard $CaSnO_3$ absorber at ambient temperature. The counting system was a scintillation counter from Harshaw, equipped with a 1 mm thick (Tl)NaI crystal. A palladium foil was used to absorb the 25.04 keV and the 25.72 keV X-ray lines generated by the source decay from the 11/2 spin level of the ^{119m}Sn precursor to the 3/2 spin of the first excited state. The Doppler velocity (± 10 mm/s) was generated by use of an Elscint driving system, including a Mössbauer MVT-4 velocity transducer, a Mössbauer MDF-N-5 waveform generator and a MFG-N-5 driver. The amplifier, the single channel analyzer and the multichannel analyzer are combined in the Tracor Northern TN7200 system. After, the data were transferred to a computer for storage and processing. Low temperature spectra were recorded using an ADP Cryogenics helium closed-cycle refrigerator equipped with a two-stage Displex®. Computer processing of the data was performed using the MOSGRAF-2009 suite [12].

4 RESULTS AND DISCUSSION

4.1 Identifying the charge carriers: Electrons or ions?

Charge transport in metallic conductors results from the motion of electrons that are weakly held to metal atoms. In solid compounds, the charge carriers can also be electrons or holes, and ions in ionic conductors, also called *superionic conductors* and *solid electrolytes* when the conductivity is similar to that in the molten state. In order for electrons to be sufficiently mobile to flow over long distances in a solid when the sample is subjected to a moderate voltage, the conducting electrons must be in a conduction band, or easily promoted to such a band. Electrons in bonding orbitals are held in place simultaneously by two atoms and therefore, they cannot move without breaking chemical bonds and destroying the material. This destructive procedure would require a voltage that is much higher.

In the MF_2 fluorides with the fluorite-type structure where M is an alkaline earth metal, i.e. other than PbF_2, the M^{2+} ions have a noble gas configuration and hence they have no valance electron, therefore they cannot contribute to electronic conductivity. The case of β-PbF_2 is different since lead is not an alkaline earth metal. It belongs to group 14, the carbon group, in period 6. The electronic structure of Pb is [Xe] $4f^{14}\ 5d^{10}\ 6s^2\ 6p^2$, hence it has four valence electrons, in agreement with its group number of 14. Ionization to the +2 oxidation number results in the loss of the $6p^2$ electrons, therefore the electronic structure of the Pb^{2+} ion is [Xe] $4f^{14}\ 5d^{10}\ 6s^2$. While the core electronic structure of the Pb^{2+} ion is [Xe] $4f^{14}\ 5d^{10}$, the $6s^2$ valence electrons are still present on the ion and, since they are not involved in bonding, they can be suspected to be sufficiently mobile and give rise to electronic conduction. Transport number measurements in a polarization cell at a constant voltage using a gold blocking electrode according to the method designed by Agrawal [13] have shown that the transport number for fluoride ions is ≥99% in all cases, including in the case of β-PbF_2.

The case of stannous fluoride SnF_2 is more surprising. It shows three solid phases: (i) α, obtained by crystallization from aqueous solutions, monoclinic, (ii) γ, obtained by heating α, stable only above the α → γ transition temperature (142–165°C, depending on particle size), tetragonal, and (iii) β, exists only below 66°C, obtained by cooling γ through a second order paraelastic → ferroelastic phase transition and is always metastable, orthorhombic [8], [14]–[17]. Bonding is strongly covalent in the three phases: α-SnF_2 has a tetrameric molecular structure while the two others form a polymeric network. Despite, the strong covalency of the bonds, the electrical conductivity of SnF_2 is similar to that of β-PbF_2 at high temperature, the conductivity of α being the highest, and the transport number for fluoride ions is above 97%, therefore the lone pair present on tin seems to not participate in the conduction mechanism [18]. The mechanism of formation of free F^- ions is not understood. In addition, the structure shows no obvious site that could give rise to Frenkel defects. In some divalent tin compounds, electrons from the lone pair are the charged mobile species. This was shown to occur in the perovskite structure $CsSnBr_3$ where the compound is a semi-conductor at ambient temperature and becomes a metal conductor at high temperature, with the transition being fully reversible on cooling [19].

We have designed a simple spectroscopic method to determine the possible role of the tin lone pair in the conduction mechanism. Three possible cases were identified and can be easily characterized by ^{119}Sn Mössbauer:

(a) The tin lone pair is stereoactive, i.e. on a hybrid orbital: Tin belongs to the same group as lead, group 14, and it has therefore four valence electrons. The electronic structure of the tin atom is [Kr] $4d^{10}$ $5s^2$ $5p^2$. In that case, bonding is covalent, the tin orbitals are hybridized sp^3 (tetrahedral electron pair geometry, three Sn-F bonds in a triangular pyramidal molecular geometry), or tin is hypervalent with five orbitals (trigonal bipyramidal electron pair geometry, see-saw molecular geometry with four bonds) or with six orbitals (octahedral electron pair geometry and square pyramidal molecular geometry with five bonds). In all cases, two of the valence electrons are used for bonding and the two others form the lone pair that occupies one of the hybrid orbital and distorts the coordination, and is said to be *stereoactive*. The above geometries are in agreement with Gillespie and Nyholm's VSEPR (Valence Shell Electron Pair Repulsion) theory [20]. The lone pair, being localized on a hybrid orbital, creates a highly distorted tin coordination that generates a large e.f.g. acting at the tin site, hence a large quadrupole splitting similar to that in Fig. 1(b). The presence of the 5s electrons, partly in bonds and partly in the lone pair, due to orbital mixing at hybridization, results in a significantly positive isomer shift. When the lone pair is stereoactive, it is locked in a hybrid orbital. Moving such electrons would destroy the hybridization, changing the bonding system and thus probably destroy its crystal structure. Therefore, if the Mössbauer spectrum is similar to that of Fig. 1(b), the lone pair electrons cannot move and therefore the compound is very unlikely to be an electronic conductor. If it is a good electric conductor, ions are likely to be the charge carriers.

The Mössbauer spectrum of α-SnF_2 (Fig. 2) and that of α-$PbSnF_4$ (Fig. 3) clearly show a large quadrupole doublet at positive isomer shifts similar to that of Fig. 1(b), characteristic of the tin lone pair being stereoactive in both cases, therefore it is locked on a hybrid orbital and cannot participate in the conduction mechanism, hence the fluoride ions must be the charge carriers.

(b) The tin lone pair is not stereoactive, i.e. it is located on the native 5s orbital: In this case, bonding is ionic. The electronic configuration of the Sn^{2+} stannous ion is [Kr] $4d^{10}$ $5s^2$. The spherical Sn^{2+} ion allows a highly symmetrical tin coordination resulting in little or no lattice e.f.g. In addition, since s orbitals are spherical, they generate no valence e.f.g. It results a total e.f.g. that is zero or near zero. This gives a single Mössbauer line. Furthermore, since there is no orbital mixing, all the 5s electron density is in the spherical orbital around the core of the Sn^{2+} ion, resulting in a higher isomer shift. In short, a single line at ca. 4 mm/s is

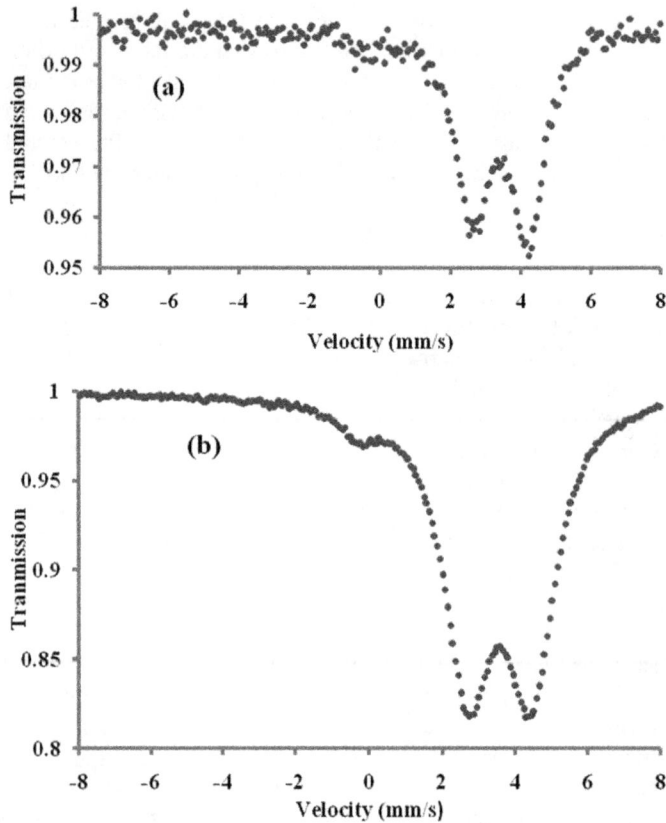

Figure 2: ^{119}Sn Mössbauer spectrum of α-SnF$_2$: (a) at ambient temperature, (b) at 12.5 K.

obtained. This is the case of CsSnBr$_3$ below the temperature where it becomes metal conductor (Fig. 1(a)).

(c) The tin lone pair moves to a conduction band, i.e. it is no longer located in the vicinity of the tin core. Tin behaves like a Sn^{4+} ion. Bonding is ionic The electronic configuration of the Sn^{4+} stannous ion is [Kr] $4d^{10}$. The spherical Sn^{4+} ion allows a highly symmetrical tin coordination resulting in little or no lattice e.f.g. In addition, since there are no more valence electrons, there is no valence e.f.g. Like in the previous case, it results a total e.f.g. that is zero or near zero. This gives a single Mössbauer line. Furthermore, since there is no more valence electron near tin, it is locally a Sn^{4+} ion, and it gives the Mössbauer spectrum of a tin (IV) compound with no quadropole splitting. This is the same spectrum as that of CaSnO$_3$, a single line at 0 mm/s (Fig. 1(c)).

4.2 How the presence of tin(II) in MSnF$_4$ enhances the conductivity by up to 10^3

Even though covalently bonded SnF$_2$ is an ionic conductor equivalent to fluorite-type β-PbF$_2$, the structure of which is ionic, it seems hard to understand how combining covalently bonded SnF$_2$ with ionic MF$_2$ (M = Pb and Ba) increases the fluoride ion conductivity by three orders of magnitude. It would seem that replacing half of the other metal by tin would replace half

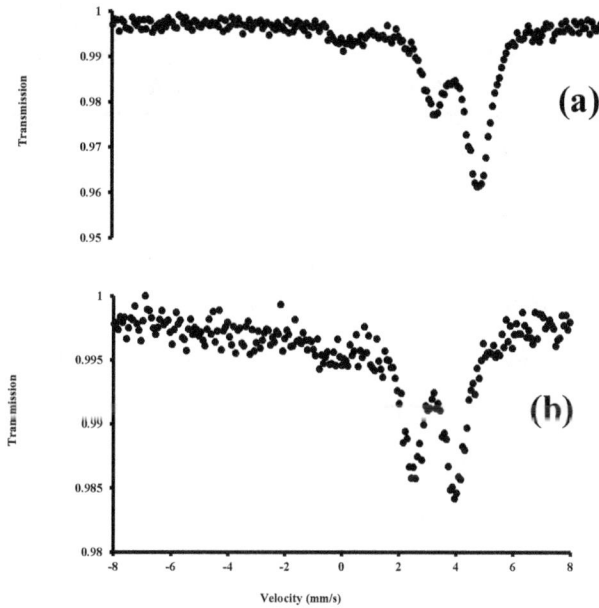

Figure 3:   Ambient temperature Mössbauer spectra of highly oriented α-PbSnF$_4$: change of the asymmetry of the quadrupole doublet with the orientation of the sample relative to the γ-ray beam direction: (a) θ = 0°, (b) θ = 45°. θ is the angle between the γ-ray beam and the normal to the plane of the sample.

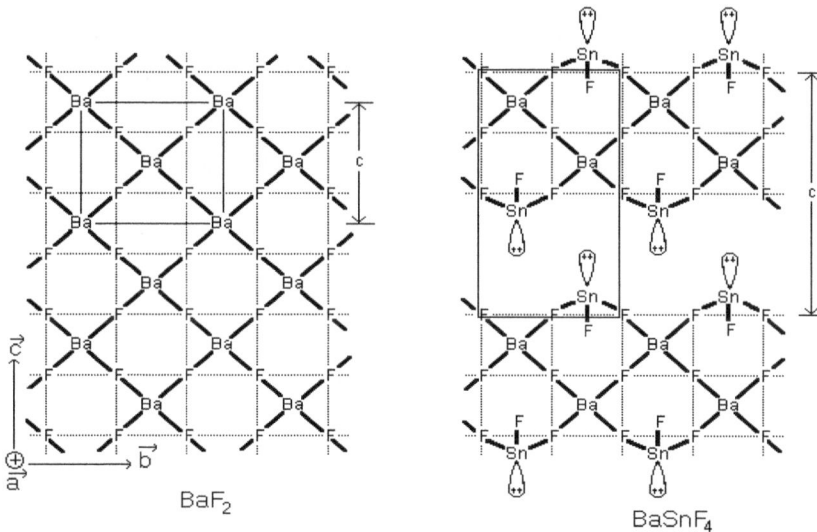

Figure 4:   Projection of a slice of the structure of BaF$_2$ and BaSnF$_4$ on the *(a,b)* plane in the BaF$_2$ axes.

of the fluoride ions by covalently bonded fluorine, hence decreasing the number of charge carriers (F^-) by 50%. This should reduce substantially the conductivity. Since it increases it instead, the answer is elsewhere. Since the number of charge carriers does not seem to be the key factor, perhaps it is the ability of the charge carriers to move. The Frenkel defect model suggests that the interstitial sites for lodging the F^- ions during the motion are located inside the $\square F_8$ cubes ($\square$ = metal ion vacancy) and the high conductivity of the fluorite-type compounds is attributed to the large number of such sites: half of the F_8 cubes have no metal ion in their center. It can be seen on Fig. 4 that for BaF_2, BaF_8 cubes and $\square F_8$ cubes alternates parallel to the b and c axes of the unit-cell, and the same happens parallel to the a axis, perpendicular to the figure, due to the cubic symmetry.

The crystal structure of $BaSnF_4$ (α-$PbSnF_4$ type) is a superstructure of that of BaF_2 by doubling the periodicity along the c axis with a. ... Ba Ba Sn Sn Ba Ba Sn Sn. ... order parallel to c. The general network or alternating MF_8 and $\square F_8$ cubes is still present within planes parallel to (a,b), it is interrupted along the c axis where the following order is observed:

[MF_8 $\square F_8$ sheet] [MF_8 $\square F_8$ sheet] [lone pair sheet] [MF_8 $\square F_8$ sheet] [MF_8 $\square F_8$ sheet].

The tin lone pair axis for all Sn atoms is parallel to c and all cluster in sheets parallel to (a,b). These sheets of lone pairs create a direction of structural weakness in the material and is responsible for the two-dimensional crystallite shape in the form of very thin sheets. This creates an enormous amount of preferred orientation in polycrystalline samples, that is responsible for the large asymmetry of the Mössbauer spectrum (Fig. 3(a)) that is considerably reduced by rotation of the sample relative to the γ-ray beam (Fig. 3(b)). On the other hand, the asymmetry of the Mössbauer spectrum of α-SnF_2 does not change with the sample orientation in the γ-ray beam. The significant asymmetry at ambient temperature (Fig. 2(a)) disappears upon cooling the sample to cryogenic temperatures (Fig. 2(b)). This temperature dependent asymmetry is due to the anisotropy of the thermal vibrations and it is called the Goldanskii–Karyagin effect [21].

How can the fluoride ions move so efficiently in the α-$PbSnF_4$ type structure? All the $\square F_8$ cubes that contain no metal ion are occupied by a fluorine atom that forms an axial bond to tin, in trans-position with the lone pair. This is a very short, very strongly covalent Sn-F bond, therefore that fluorine atom cannot be mobile. In addition, that axially bonded fluorine atom leaves no room to lodge interstitial fluoride ions in the $\square F_8$ cubes. I seems therefore that all the potential interstitial sites present in the fluorite type MF_2 have disappeared in the $MSnF_4$ structure. The mobile fluoride ions must be those bonded to barium only since any fluorine atom bonded to tin forms a Sn-F covalent bond. The only room available for the motion of the fluoride ions is in the sheets of lone pairs. The presence of the lone pairs will prevent any bonding to tin and the repulsions between the charge of the lone pair and that of the F^- ions, both negative, will result in no attraction of the fluoride ions in this space and therefore they will be highly mobile.

5 CONCLUSION

In this study, the mechanisms of electrical conduction in ionic solids were analyzed with emphasis on fluorides with the fluorite-type structure and related $MSnF_4$. The method of determining by ^{119}Sn Mössbauer spectroscopy whether the tin(II) lone pair can be conducting or not was described. The possible pathways for the long distance motion of fluoride ions in both structures were evaluated. A particular question, why covalently bonded SnF_2 increases the ionic conductivity of the MF_2 in $MSnF_4$, was studied in details.

ACKNOWLEDGEMENT

Financial support in the early part of this work from the Natural Science and Engineering Research Council of Canada and from Concordia University is acknowledged. Technical support from Concordia University is much appreciated.

REFERENCES

[1] Faraday, M., Experimental researches in electricity. *PhilosTrans R. Soc. London*, **128**, pp. 1–40, 1838. DOI: 10.1098/rstl.1838.0008.

[2] Schoonman, J., Retarded ionic motion in fluorites. *Solid State Ionics*, **1**, pp. 121–131, 1980.

[3] Chadwick, A.V., Diffusion in fast-ion conductors. *Diffusion in Materials. NATO ASI Series (Series E: Applied Sciences)*, vol. 179, eds A.L. Laskar, J.L. Bocquet, G. Brebec, C. Monty, Springer: Dordrecht. DOI: 10.1007/978-94-009-1976-1_24.

[4] Dénès, G., Failure of the Frenkel defect model to explain the trend in ionic conductivity in the MF_2 fluorite structure and related $MSnF_4$ materials. *Mat. Res. Soc., Proc.*, **369**, pp. 295–300.

[5] Dénès, G., Birchall, T., Sayer, M. & Bell, M.F., $BaSnF_4$ – A new fluoride ionic conductor with the α-$PbSnF_4$ structure. *Solid State Ionics*, **13**, pp. 213–219, 1984.

[6] Yang, L. et al., SnF_2-based fluoride ion electrolytes $MSnF_4$ (M = Ba, Pb) for the application of room-temperature solid-state fluoride ion batteries. *J. Alloys Comp*, **819**, p. 152983, 2019.

[7] Mori, K., Mineshige, A., Emoto, T. & Fukunaga, T., Electrochemical, thermal, and structural features of BaF_2 –SnF_2 fluoride-Ion electrolytes. *J. Phys. Chem. C.* DOI: 10.1021/acs.jpcc.1c03326.

[8] Dénès, G., About SnF_2 stannous fluoride. *VI. Phase Transitions Mat. Res. Bull.*, **15**, pp. 807–819, 1980.

[9] Weast, R.C. & Astle, M.J. (eds), *CRC Handbook of Chemistry and Physics*, 61st edn., CRC Press: Boca Raton, FL, p. B-80, 1980–1981.

[10] Dénès, G., The "Bent copper tube": A new inexpensive and convenient reactor for fluorides of metals in suboxidation states. *J. Solid State Chem.*, **77**, pp. 54–59, 1988.

[11] Dénès, G., Pannetier, J. & Lucas, J., Les fluorures à structure PbFCl (M = Pb, Sr, Ba). *C. R. Acad. Sc. Paris*, **280C**, pp. 831–834, 1975.

[12] Ruebenbauer, K. & Duraj, Ł., www.elektron.up.krakow.pl/mosgraf-2009.

[13] Agrawal, R.C., DC Polarisation: An experimental tool in the study of ionic conductors. *Indian J. Pure Appl. Phys.*, **37**, pp. 294–301, 1999.

[14] Dénès, G., Pannetier, J. & Lucas, J., About SnF_2 stannous fluoride. I. Crystallochemistry of α-SnF_2. *J. Solid State Chem.*, **30**, pp. 335–342, 1979.

[15] Dénès, G., Pannetier, J. & Lucas, J., About SnF_2 stannous fluoride. I. Crystal structure of β- and γ-SnF_2. *J. Solid State Chem.*, **30**, pp. 1–11, 1980.

[16] Pannetier, J., Dénès, G., Durand, M. & Buevoz, J.L., $\beta \rightleftarrows \gamma$-$SnF_2$ phase transition: Neutron diffraction and NMR study. *J. Physique*, **41**, pp. 1019–1024, 1980.

[17] Dénès, G., About SnF_2 stannous fluoride. IV. Kinetics of the $\alpha \rightarrow \gamma$ and $\beta, \gamma \rightarrow \alpha$ transitions. *J. Solid State Chem.*, **37**, pp. 16–23, 1981.

[18] Ansel, D., Debuigne, J. Dénès, G., Pannetier, J. & Lucas, J., About SnF_2 stannous fluoride. V. Conduction characteristics. *Ber. Bunsenges. Phys. Chem*, **82**, pp. 376–380.

[19] Donaldson, J.D., Silver, J., Hadjiminolis, S. & Ross, S.D., Effects of the presence of valence-shell non-bonding electron pairs on the properties and structures of cesium tin(II) bromides and of related antimony and tellurium compounds. *J. Chem. Soc. (A)*, pp. 666–672, 1972.

[20] Gillespie, R.J. & Nyholm, R.S., Inorganic stereochemistry. *Quart. Rev. Chem. Soc*, **11**, pp. 339–380, 1957.

[21] Birchall, T., Dénès, G., Ruebenbauer, K. & Pannetier, J., Goldanskii-Karyagin effect in α-SnF$_2$: A neutron diffraction and Mössbauer absorption study. *Hyperf. Inter.*, **30**, pp. 167–183, 1986.

Author index

WITPRESS ...for scientists by scientists

Linear and Non-linear Continuum Solid Mechanics

S. HERNÁNDEZ, *University of A Coruña, Spain and* **A. N. FONTAN**, *University of A Coruña, Spain*

Deformable solids, that is to say, those which undergo changes in geometry when subjected to external loads or other types of solicitations, as well as other related topics are the focus of this book.

Within the main field, this text deals with advanced linear elasticity and plasticity approaches and the behavioural study of more complex types of materials. This includes composites of more recent manufacture and others whose material characterisation has only recently been possible. It also describes how linear elastic behaviour extends to anisotropic materials in general and how deformations can result in small or large strain components. The information on plastic behaviour expands to include strain hardening of the materials.

Amongst other new topics incorporated into this volume are studies of hyperelastic materials, which can represent elastomeric and some types of biological materials. A section of the book deals with viscoelastic materials, i.e. those who deform when subjected to long-term loads. The behaviour of viscoplasticity, as well as elasto-viscoplasticity, describes well other types of materials, including those present in many geotechnical sites.

The objective of this volume is to present material that can be used for teaching continuum mechanics to students of mechanical, civil or aeronautical engineering. In order to understand the contents the reader only needs to know linear algebra and differential calculus.

Examples have been included throughout the text and at the end of each chapter, exercises are presented which can be used to check on comprehension of the theoretical information presented.

ISBN: 978-1-78466-271-4   **eISBN: 978-1-78466-272-1**
Published 2021 / 206pp

www.ingramcontent.com/pod-product-compliance
Lightning Source LLC
Chambersburg PA
CBHW062006190326
41458CB00009B/2982

9 781784 664374